DR. JOE AND WHAT
YOU DIDN'T KNOW

PRAISE FOR *THE GENIE IN THE BOTTLE*:

"Often lighthearted, other times deadly serious, he covers a range of topics, 64 in all, that includes tap water quality, methane, bag balm, and even 'flubber.' No doubt readers will find a lot of stuff here about a lot of things they have never even considered."

— Ted Hainworth, Star Phoenix, 29 December 2001

PRAISE FOR *RADAR, HULA HOOPS, AND PLAYFUL PIGS*:

"Joe Schwarcz tells it like it is. Whether he's plumbing the mysteries of chicken soup or tracing the development of polyethylene, Schwarcz takes a little history, adds a dash of chemistry, and produces a gem of an essay every time. I wish he'd been my chemistry professor when I was in school."

— Christine Gorman, senior writer, Time magazine

"Dr. Schwarcz has written a book that has done three things which are difficult to do. First, the book is enormously enjoyable — it commands and holds your attention. Second, it explains science and scientific phenomena in a simple and yet accurate way. And third, it stimulates you to think logically and in so doing, it will lead to a scientifically literate reader who will not be so easily misled by those who wish to paint science and technology as being a danger to humankind and the world around us."

— Michael Smith, Nobel Laureate

"It is hard to believe that anyone could be drawn to such a dull and smelly subject as chemistry — until, that is, one picks up Joe Schwarcz's book and is reminded that with every breath and feeling one is experiencing chemistry. Falling in love, we all know, is a matter of the right chemistry. Schwarcz gets his chemistry right, and hooks his readers."

— John C. Polanyi, Nobel Laureate

DR. JOE AND WHAT
YOU DIDN'T KNOW

DR. JOE SCHWARCZ

Director
McGill University Office for Science and Society

ECW PRESS

Published by ECW PRESS
665 Gerrard Street East, Toronto, Ontario, Canada M4M 1Y2

NATIONAL LIBRARY OF CANADA CATALOGUING IN PUBLICATION DATA

Schwarcz, Joseph A.
Dr. Joe and what you didn't know: 177 fascinating questions and answers
about the chemistry of everyday life / Joe Schwarcz.

Includes index.

ISBN 978-1-55022-577-8

ALSO ISSUED AS:
978-1-55490-577-5 (PDF)
978-1-55490-538-6 (EPUB)

1. Chemistry—Popular works. I. Title.

QD37.S374 2003   540   C2003-902203-X

Copy editor: Mary Williams
Cover design: Guylaine Régimbald—Solo Design
Cover illustration: Peter Till/Getty Images
Interior design and typesetting: Yolande Martel
Interior illustrations: Brian Gable
Author photo: Tony Laurinaitis
Production: Emma McKay

This book is set in Stempel Garamond and Koch Antiqua.

The publication of *Dr. Joe and What You Didn't Know* has been generously supported by the Canada Council, the Ontario Arts Council, and the Government of
Canada through the Book Publishing Industry Development Program.

Canadä

PRINTED AND BOUND IN CANADA

ECW PRESS
ecwpress.com

INTRODUCTION

What does being barefoot and pregnant on a cactus have to do with cherry ice cream? You probably think that's a pretty bizarre question. But there is method to the madness. For over two decades now, I've tried to answer the public's questions about everyday chemistry on the radio, I hope with some degree of success. Then, a few years ago, I had an idea. Why not spice things up by turning the situation around and asking my audience questions?

Everybody seems to like quiz shows, especially when there are prizes at stake, so I took to beginning each program with a question, offering a prize for the first right answer. We're talking local radio here, not national TV, so the prizes were modest, but we did manage to come up with some books, gift certificates, restaurant meals, and highly coveted "Dr. Joe" T-shirts. A new adventure was in the offing! I started off with some pretty straightforward questions — like, "Why does ouzo turn white when you add water to it?"; and, "What is emu oil?"; and, "Why does lemon juice lighten the color of tea?" All good questions, I thought, and somewhat challenging.

Indeed, for the first few weeks, callers struggled to answer my questions; then suddenly it seemed as though they had taken smart pills. They started providing the correct answers with impressive frequency. It didn't take long for me to figure out what was going on. I wasn't testing the audience anymore, I was posing my questions to Google. And Google, as we know, is pretty smart. As soon as a question passed my lips, those little fingers hit the keyboard. I tried instituting an honor code, requesting that respondents not run to their computers. Yeah, sure.

Obviously, another approach was needed. I attempted to phrase the questions in ways that would confound the search engines. Like, "What is the connection between tangerine trees, marmalade skies, and morning glory?"; or, "What is the link between Frankenstein and frog legs?" This approach seemed to work, and that's why you'll find a number of unusual-sounding questions in this little volume, which is a collection of the questions I've posed on my radio show over the years. Of course, you'll also find the answers — at least, my version of them.

The questions come from a number of fields, but all have some interesting scientific connection. There is no specific order to the questions, no systematic attempt to educate. What I've tried to do is demonstrate science's broad scope and show how scientific pursuit links to so many areas of our culture. So, come along for the ride, poke around, read a little here, a little there, pick up some bits of knowledge to fling around the dinner table, and have some fun doing it. And if you want to know what being barefoot and pregnant on a cactus has to do with cherry ice cream, just turn the page.

1. How would you relate "barefoot, pregnant, and on a cactus" to cherry or strawberry ice cream?

It all comes down to the fascinating little insect called *dactylopius* coccus.

When Hernán Cortéz arrived in Mexico in 1518, he was intrigued by the beautifully colored Aztec fabrics he saw there. The source of the dye appeared to be seeds on the surface of certain cactus plants, but closer scrutiny revealed that they were not seeds at all. They were little bugs. Today, we know these insects as cochineal and the dye they yield as carmine. Montezuma, the Aztec king, was so fond of wearing robes made of carmine-dyed fabric that he imposed a tax upon his subjects to be paid in dried cochineal insects.

The pregnant female cochineal bug produces the brilliant red dye that became the first product ever exported from the New World to the Old. Soon, Europeans were dying their wool and silk with the insect extract. Maybe the most memorable use of cochineal was the bright scarlets for which the Gobelin tapestries of Paris became famous.

Producing the dye is not an easy business. The female insects, which feed on the red cactus berries and concentrate the dye

in their bodies and in their larvae, are scraped off the cactus and dumped into hot water, where they instantly die. They are then dried in the sun and crushed into a powder, which is added to water or to a water-alcohol mixture. For fabrics, a mordant, such as alum, which binds the color to the material, is generally used. Carminic acid, the active coloring agent, is one of the safest existing dyes, and it is commonly used in foods and cosmetics. Candies, ice cream, beverages, yogurt, lipstick, and eye shadow can all be colored with cochineal.

Allergic reactions to the dye are rare. There have been reports of people reacting to the aperitif Campari, pink popsicles, maraschino cherries, and red lipstick, but more people suffer reactions to other food and cosmetic ingredients. In one instance, the face of a little boy who was kissed by his loving grandmother became swollen. It seems he had been sensitized to carmine, probably through food or candy, and he had reacted to the coloring in her lipstick. When reactions do occur, they tend to be in the form of hives and swelling, although one anaphylactic reaction to Campari-Orange has been reported.

Cochineal insects are very small, so it takes about seventy thousand females to produce a pound of dye. The males are quite useless in this regard. Like the males of most species, they are duller than the females. They are also rare and live for only a week; during their lifetime, they mate with as many females as possible. (Maybe they are not so dull after all.) So, how do the dye makers separate the sexes? Well, the males can fly, but the wingless females cannot. When the cactus is disturbed, the males scoot, but the females cannot escape. They are scraped off, destined to color some of our cherry or strawberry ice cream. I know that many of you may not find the prospect of ice cream colored with bug juice appetizing, but it is an effective and safe dye. And, of course, it's all natural.

2. What condition would you have if you were being treated with carbamide peroxide?

———

Stained teeth. Carbamide peroxide is the active ingredient in most tooth-whitening products, and it works by releasing hydrogen peroxide, which in turn yields hydroxyl free radicals, which can break down colored molecules.

Hydrogen peroxide itself is a liquid and difficult to apply to teeth, but when it's mixed with urea it forms a gel of carbamide peroxide that can easily be painted on teeth, placed into trays fitted to the teeth, or incorporated into whitening strips. Thickeners such as carbopol and glycerin are often used to achieve the right consistency.

Tooth discoloration is mostly the result of colored substances in foods and drinks that embed themselves over time in the calcium phosphate that makes up the tooth's outer coating, the enamel. Tannins in tea and coffee, anthocyanins in blueberries, and polyphenols in red wine are just some of the compounds that can discolor teeth. A further complication is that dentin, the mix of proteins and calcium phosphate that lies beneath the enamel, yellows naturally with age. The molecules responsible for tooth discoloration tend to have a network of carbon-carbon double bonds. Such unsaturated systems, as they are called, absorb some colors but reflect yellow. Hydroxyl radicals are highly reactive and can disrupt these double bonds, leading to whitened teeth.

Applying various peroxide products to the teeth is generally quite a safe and simple procedure, although some people experience heightened sensitivity to cold after their dentists apply products containing high concentrations of hydrogen peroxide. Products developed for home use generally contain only 3 to 6 percent hydrogen peroxide and do not cause sensitivity, but they may take weeks to lighten discolored teeth. We may not yet have an ideal system for treating stained teeth, but carbamide peroxide is surely a great improvement over historical methods, which included swirling with urine or rubbing the teeth with a mixture of chalk and ground rabbit skull.

3. By law, chlorofluorocarbons (cfcs) used as refrigerants in refrigerators must be removed before the appliances are discarded. This solves only part of the ozone-depletion and global-warming problem attributed to refrigerators. Why?

―――

The walls of refrigerators have to be heavily insulated to ensure efficient cooling. Typically, polyurethane foam insulation has been used for this purpose, and guess what it used to be blown with? Chlorofluorocarbons!

Foams are created by blowing a gas into a material to form bubbles. Of course, the gas must not react with the material, and, in the case of insulation, it should not transmit heat readily. CFCs, the substances already used as refrigerants, seemed ideal — at least until their environmental consequences were discovered. Legislation was then introduced calling for the removal of the refrigerant from all discarded refrigerators.

Most people would be surprised to learn that a far greater quantity of CFCs was used to make foam insulation for fridges than was used for refrigeration. A typical fridge may have a couple of hundred grams of refrigerant, but it can hold twice as much blowing agent captive in its insulation. And "captive" is the appropriate expression, because studies have shown that more than 90 percent of the original blowing agent is still present in a refrigerator fifteen years after it has been discarded.

Unless special methods are employed, the blowing agent is released into the atmosphere when such fridges are recycled for their metal content. Shredding the fridge into small pieces in an airtight chamber allows for recovery of CFCs. This technique is expensive, but it can have huge environmental benefits. Refrigerators manufactured these days do not present this problem. They contain cyclopentane as the insulating gas, and this has no effect on ozone depletion and a negligible effect on global warming.

CFCs as refrigerants were replaced in the 1990s by HFCs (hydrofluorocarbons), which do not damage the ozone layer but do contribute to the greenhouse effect. Some manufacturers are now switching to isobutane as a refrigerant, because, like cyclopentane, it has a minimal impact on the environment. Given that millions and millions of discarded fridges are piled up around the world, the problem associated with the CFC content of polyurethane foam insulation is not a trivial one.

4. Why does the common symbol for medicine depict a snake?

A snake coiled around a staff is widely recognized as a symbol of healing. The staff belongs to Asklepios, the Greek god of medicine.

In ancient Greece, the sick would go to shrines called *asklepieia*, where priests would conduct healing ceremonies, often using sacred serpents. We don't know whether the snakes actually had a practical function in the treatment of disease or whether they just scared people into feeling better, but Italian researchers have examined the healing potential of the "four-lined snake," which is commonly found in Greece. An ancient relief depicting a wounded boy and the mouth of a large snake is what prompted the research. It turns out that snake saliva contains epidermal growth factors, which really may help heal wounds. Perhaps snakes are blessed with this chemical because their mouths are vulnerable to damage as they ingest their prey.

Sacred dogs were also kept in the *asklepieia*. Was it their job to lick wounds? There actually is some evidence that dog saliva, like that of snakes, contains epidermal growth factors. These substances induce healing by causing the proliferation of certain skin cells. Maybe that's why dogs are always licking themselves!

And what happened to Asklepios in Greek mythology? According to the story, the god of medicine was slain by Zeus because he feared that Asklepios would make all men immortal. But such notions were overturned by Hippocrates, who made the revolutionary suggestion that diseases were not caused by the gods and could not be cured by them.

Hippocrates initiated a process of careful observation and experimentation. He separated myth and magic from rational therapy. "Every natural event has a natural cause," he maintained. Hippocrates investigated symptoms and was able to predict the course of disease. But Asklepios's reliance on the healing power of snakes may yet turn out to have some merit. Proteins isolated from certain snake venoms have powerful anticlotting effects on the blood and may one day be used in the treatment of thrombosis.

5. In the late 1800s, fashionable ladies accentuated their derrieres by wearing bustles under their skirts. To further emphasize their protruding rear ends, many would bend forward as they walked, assuming a posture that came to be known as the "Grecian bend." What does this have to do with the construction of the Brooklyn Bridge?

The workers who built the underwater foundations of the Brooklyn Bridge often experienced excruciating pain when they returned to the surface of the Hudson River. It caused them to double over, a little like the bustle-wearing women with their "Grecian bends."

Decompression sickness was what afflicted these workers, but they referred to it as "the bends." The gigantic pylons that support the bridge had to be constructed deep in the riverbed, and the construction workers labored in large, open-bottomed timber chambers, or caissons, on the floor of the Hudson. Inside these caissons, they toiled away, excavating soil and rock. The surrounding water exerted tremendous pressure on the chamber walls, so the air inside had to be pressurized to prevent the caissons from collapsing.

Since the extent to which a gas dissolves in a liquid is determined by the pressure exerted by the gas on the surface of the liquid (Henry's Law), at high pressures, more nitrogen (which makes up 80 percent of air) dissolves in blood. If the pressure is released too quickly, as it was when the bridge workers rose to the river surface, the nitrogen comes bubbling out of solution and causes the bends.

The risks of working in a chamber of compressed air at the bottom of a river were little understood in the late 1800s. Even the chief engineer of the bridge, Washington A. Roebling, didn't appreciate the severity of the problem. In 1872, after spending twelve hours breathing pressurized air in a submerged caisson, he lost consciousness and became permanently paralyzed from the waist down. Over a hundred other bridge workers were afflicted by the bends, and three died.

The same problem plagued the builders of the Holland Tunnel — the first subway tunnel under the Hudson — until E. W. Moir installed decompression chambers at the work site. Moir realized that a victim of the bends could be treated by being placed inside a high-pressure chamber. There he would remain until the nitrogen in his body was forced back into solution in the blood, to be released at a controlled rate — a slow decompression. By the time the Holland Tunnel was completed, in the 1920s, the situation was well in hand, and not a single worker died from the bends. The tunnel was designed so that workers had to pass through decompression chambers, and those working under high pressure were allowed to work only for short periods.

Robert Boyle, perhaps the greatest scientist of the seventeenth century, would certainly have appreciated this. It was he who noted that rapid decompression can cause previously dissolved gases to come out of solution. How did he prove it? He placed a snake inside a chamber, reduced the pressure, and observed a gas bubble forming in the snake's eye. Gas studies like this one led him to formulate Boyle's Law, which states that the volume of a gas is inversely proportional to pressure. If you want a demonstration of this law, blow up a small balloon and take it along on your next airplane ride.

6. What happens to the alcohol when wine changes into vinegar?

Simple. It disappears, because it is the alcohol that gets converted to vinegar. But even simple answers like this one have interesting stories behind them.

When the alcohol in wine changes into vinegar, there are two processes involved. The first one is relatively minor. Ethanol, the alcohol of beverages, reacts with oxygen to form acetic acid, a dilute solution of which we refer to as vinegar. This happens only to a very small extent, because the wine doesn't come into contact with much oxygen. What really causes wine to turn to vinegar is contamination with a bacteria called *Acetobacter aceti.*

This very common bacterium produces an enzyme that converts ethanol to acetic acid. It can be found on the grapes used to make wine, but the most typical source of contamination is the fruit fly. That's why vintners take such elaborate measures to keep the little bugs out of their fermenting mixtures. Once *Acetobacter* bacteria get a foothold, they begin to multiply and soon form a cellulose-based, jelly-like substance called mother of vinegar. In the Philippines, this substance is regarded as a delicacy. A traditional Philippine sweet, called *nata de coco* or *nata de pina*, is made by mixing the bacterial cellulose with sugar.

In general, the conversion of alcohol in wine to acetic acid is considered undesirable. But not always. Wine vinegar is a popular gourmet grocery item. It's made by introducing mother of vinegar into wine to encourage the production of acetic acid. Many people prefer wine vinegar to regular vinegar because, in addition to acetic acid, it has numerous flavor compounds that were produced by the original fermentation.

It is possible, however, to make vinegar without using wine. Ethanol can be made from ethylene, which in turn is made from petroleum. The ethanol can then be converted to acetic acid by reaction with oxygen. Large amounts of acetic acid are produced industrially in this fashion. Diluting the pure acetic acid in water to a concentration of 5 percent produces vinegar. And if all you plan to do with the stuff is clean your boots or sprinkle it on your french fries, then it's good enough. But if you're having guests for dinner and serving up a salad, spring for the wine vinegar. And, for dessert, why not offer some *nata de coco*?

7. What are gel candles, and are they really dangerous?

All kinds of horror stories travel around the Internet — such as the one about gel candles that explode and burn down your house. These stories are usually buttressed by the accounts of those who have "seen it happen." Well, the gel candle story is almost 100 percent bunk.

These candles have become very popular because they're pretty and they burn much longer than regular candles. Candle makers can also incorporate a diversity of fragrances and dyes into their gel products. The typical gel candle purchaser probably doesn't know that the candle's origins can be traced back over 1,300 years to something historians have referred to as "Greek fire" — which wasn't actually invented by the Greeks. This early version of a flamethrower was first used by the defenders of Constantinople in the seventh century; with a primitive pump, they sprayed hot oil — made sticky by the addition of tree resins — through a tube.

Gel candle manufacturers convert purified mineral oil, a petroleum distillate product, into a gel. In order to do that, they need a substance that has thickening properties and can be mixed with the hydrocarbons that make up the mineral oil. Tree resin was a good choice when nothing else was available, but modern chemistry has provided us with far better gelling agents — such as styrene-ethylene/styrene-butadiene block copolymers, which, when heated with oil, produce a gel. Just insert a wick, and you have a gel candle.

Contrary to the Internet rumors, gel candles do not produce a mysterious explosive gas that can blow your house to smithereens. As with any burning hydrocarbon mixture, the gases produced are carbon dioxide and water vapor. Not exactly chemicals from hell. So, gel candles do not explode. Why, then, should we have any reservations about dismissing the gel candle bomb stories as totally absurd? Because gel candles really are potentially more dangerous than regular candles. That's because manufacturers pour the gel into the candle's glass container, and when the candle burns the glass heats up. If the glass isn't tempered, then it can break, and there's a remote possibility that the burning oil will spread and cause a fire.

So, the moral of the story is this: Make sure that you position your gel candles well away from combustible matter and never leave them unattended. They won't explode, but remember that Greek fire warded off many an invader and caused a lot of damage, even though it never caused anything to blow up.

8. In 1995, researchers from Cambridge University asked the Manchester Literary and Philosophical Society for a sample of an eyeball that had been sitting in a jar on a shelf since 1844. Whose eyeball was it, and what did the researchers want to do with it?

The eye in the jar made some of the most important scientific observations in history. It — and another just like it — belonged to John Dalton, the English schoolteacher who, in the late years of the eighteenth century, formulated the atomic theory.

Dalton had inferred from the way that elements combined with each other that these fundamental building blocks of matter were made of atoms and that the atoms of any element were identical to each other but different in mass from the atoms of other elements. He also meticulously recorded his observations of weather patterns, the northern lights, and the behavior of gases. As well, he discovered that he'd made these observations with eyes that were different from others: he was color-blind! Dalton already suspected that he had vision problems, because his fellow Quakers would occasionally object to the loud colors he wore; to his eye, the shade of his attire seemed quite sedate.

Then one night in 1792, Dalton noticed that a geranium plant that had appeared blue in the sunlight changed color by candlelight. (Candlelight is composed of a different range of wavelengths, or colors, than sunlight is. Newton had demonstrated this long before by passing light through a prism.) Dalton questioned his friends about it, but they were puzzled because they witnessed no such color change. Something interesting was going on.

Dalton surmised that his eyes were somehow filtering out certain colors, and this prompted him to consider the possibility that the vitreous humor, the thick liquid inside his eyeballs, was a different color from that found in the eyeballs of others. He wasn't keen on having his eyes taken out while he was still alive, but he requested that they be removed after his death and studied. His assistant, Joseph Ransome, complied. He squeezed out the liquid and found it to be perfectly normal. Then he made a hole in the back of one eyeball and looked through it. Not noticing any filtering effect, he concluded that color blindness did not stem from a physical change in the eyeball.

Ransome was wrong about that, but he wouldn't have had the means at the time to determine the cause of color blindness. Today, we can relate color blindness to malfunctioning cells in the retina, the light-sensitive layer of tissue that lines the back and sides of the eyeball. There, color is perceived by cells called cones. We have three types: one type is sensitive to blue, another to green, and a third to yellows and reds. Color blindness is a malfunction in one or more of these cell types. "Deuteropes," for example, cannot see the green part of the spectrum, "protanopes" are insensitive to red, and "tritanopes" are blind to blue.

Color vision and the problems associated with it are encoded in our genes. That's why the Cambridge researchers asked to investigate Dalton's eyeballs. By 1995, the polymerase chain reaction (PCR) had been developed to the extent that a tiny sample of DNA could be reproduced and samples large enough for laboratory investigation generated. The researchers subjected the DNA they extracted from cells in a sample of Dalton's retina to such a study and discovered that John Dalton was indeed a deuterope who saw the world differently from others. Dalton himself had presaged such genetic analysis by observing that, while his friends saw no difference in the color of the geranium by candlelight, his brother did.

Curiously, it was not the English but the French who commemorated Dalton's observations about color blindness in a significant way. The French term for color blindness is *daltonisme*.

9. What potentially dangerous compound is less likely to form in thick french fries than in thin ones?

———

Acrylamide. In April of 2002, Swedish scientists gave us yet another reason to worry about what we put in our mouths when they discovered significant amounts of acrylamide in many fried and baked foods.

The amounts were far in excess of 0.5 parts per billion, which is the World Health Organization's limit for drinking water. Polyacrylamide is used as a filtering aid in some municipal water-treatment plants, and small amounts of residual acrylamide are found in the water they process. The WHO limit was instituted because acrylamide is a known animal carcinogen.

Understandably, many people became alarmed when the Swedish researchers found thousands of parts per billion of acrylamide in chips, fries, and — horrors! — Swedish crispbread. Where was it coming from? The research community shifted into high gear, and within a couple of months we had the answer. Certain amino acids, asparagine in particular, react with glucose or sucrose under frying or baking conditions to yield acrylamide. The temperature is critical. No acrylamide forms under 120°C (248°F), and only a moderate amount forms up to 175°C (347°F). But then there is a huge escalation. In one test, fries fried at 175°C (347°F) were found to have 300 parts per billion; at 180°C (356°F), the level soared to 1,100 parts per billion. Some potato chips were found to contain as much as 3,700 parts per billion. It doesn't matter whether the chips in question are organic, either — some of the highest levels were found in chips sold in health food stores. If acrylamide is as carcinogenic in humans as it is in animals, then its presence in the food supply could be responsible for several thousand cases of cancer a year in North America.

While we can't completely avoid consuming acrylamide, we can reduce levels in food. A serving of thin fries has a greater collective surface area than a serving of thick ones, so more of the potato is exposed to high temperatures. Soaking the fries in water for an hour before frying helps, because some of the sugar in the potato will dissolve in the water. But the key to reducing acrylamide is temperature control. If we maintain the temperature of the frying oil below 175°C (347°F), then we can reduce the levels dramatically.

You can imagine the feverish activity now going on in the test kitchens of major fast-food chains as food researchers try to reduce the acrylamide content of fries. Of course, we should remember that the high fat content of french fries is still a better reason to curb our intake than acrylamide content. We can also take some comfort in a joint study conducted by the Harvard School of Public Health and the Karolinska Institute in Sweden, which found no link between the consumption of acrylamide and the occurrence of colon, bladder, or kidney cancers.

The study's researchers, who reported their results in the *British Journal of Cancer* in 2003, performed what is known as a case-control study. They examined the dietary intake of acrylamide among 987 cancer patients and compared it to that of 538 healthy people to see if they could find a link between the disease and the chemical. No such link was apparent — the cancer patients had consumed no more acrylamide than had the healthy subjects. In fact, they associated higher levels of acrylamide in the diet with a lower, not higher, risk of colon cancer. Still, we are not yet ready to declare acrylamide an anticarcinogen.

10. How does a Fizz Keeper keep the fizz in soft drinks?

It doesn't. But it is a clever bit of marketing. And who knows, the manufacturer may actually think it works because the idea behind it does seem to make sense.

The Fizz Keeper is a little hand pump that you can screw into the neck of an opened bottle in order to pressurize the contents and preserve the carbonation. Pumping air into the bottle can certainly restore the pressure above the solution, making the bottle feel hard, like it was when it was purchased. But the Fizz Keeper's manufacturers seem to be unaware of Henry's Law.

William Henry was the English chemist (1775–1836) who noted that the solubility of a gas in a liquid is proportional to the pressure of the gas above the solution (see p. 15). The presence of other gases does not matter. Consider a carbonated beverage. Before the bottle is sealed, it is pressurized with carbon dioxide. The pressure of the carbon dioxide is very high — far higher than atmospheric pressure — so a great deal of carbon dioxide dissolves. When you open the bottle, the pressurized gas escapes, and the only carbon dioxide sitting over the liquid is the atmospheric carbon dioxide, which has a tiny, partial atmospheric pressure of 0.0003. The excess carbon dioxide comes out of solution, producing the fizz. The only way to prevent that loss of dissolved carbon dioxide once you've opened the bottle is to pressurize the contents with carbon dioxide, not air.

11. What is a molecule?

If you're not quite sure, you're not alone. Surveys have shown that a mere 10 percent of the population understands what a molecule is. By contrast, 50 percent knows that the Earth goes around the sun once a year, and 40 percent realizes that electrons are smaller than atoms. This lack of familiarity with molecules is distressing, because everything in the physical world depends upon molecular action.

Molecules are the fundamental components of matter. They are made up of atoms, which in turn are composed of even smaller particles called protons, neutrons, and electrons. Perhaps it isn't really surprising that people have misconceptions about molecules, since molecules are almost inconceivably small — much too small to be seen. So, how do we know that they exist? The simplest answer may be that our ideas about molecules must be correct because we can predict and explain the behavior of matter based on the concept that everything is made up of molecules. Examples range from explanations for why taking an antacid helps heartburn to why adding an acid to milk causes it to curdle.

We chemists spend our lives thinking about, and working with, molecules. That's why we become so irritated when we are confronted with molecular silliness. Like the statement made by a meteorologist to the effect that fog is air saturated with water, and as the air cools the water molecules get bigger and bigger until they become visible. Water molecules do not change in size. They can cluster and form a liquid or separate and form a vapor. Fog does not consist of large water molecules; it consists of water molecules that have clustered to form droplets of liquid.

Even more outrageous is the claim made for the Laundry Disk, a product that is supposed to enable us to wash our clothes without detergent. Its makers say that it contains "an activated ceramic that makes water molecules smaller and enhances their ability to penetrate the fabric." The Laundry Disk does not work this way or indeed in any other way.

Water molecules do not change in size, but they do allow other molecules to squeeze between them. This is why water is such an excellent solvent. Have you ever wondered where sugar goes when you dissolve it in a glass of water? Another interesting demonstration involves a phenomenon that was first noted by the ancient Greeks. If we combine equal amounts of water and alcohol, then we get a volume that is less than the sum of the two. How come? The only possible answer is that the alcohol molecules have lodged themselves in the spaces between the water molecules. And what can you do with that mixture of water and alcohol? Drink a toast to the molecule!

12. What does sniffing chocolate fragrance have to do with losing weight?

How would you like to lose weight without dieting? All you have to do is inhale a certain odor when you get hungry. This may sound like just another diet scam, but it could actually have a scientific basis.

The idea is based on research carried out by Dr. Alan Hirsch, a scientist at the Smell and Taste Research Foundation of Chicago. Hirsch conducted a six-month-long study involving 3,193 people who were at least ten pounds overweight. Subjects were each given a vial and told not to reach for food whenever they felt hungry but to sniff the vial's contents instead. So the subjects happily sniffed — some of them up to 285 times a day.

Hirsch's subjects lost an average of five pounds a month.

He and his researchers experimented with different smells, but it didn't seem to matter — banana, apple, and peppermint yielded similar results. Hirsch's findings are still considered preliminary, because the study has not been independently reproduced, but that hasn't stopped entrepreneurs from marketing sniff-and-lose-weight products. Of course, if the smell method truly works, then all you should have to do is sniff your favorite food whenever you feel hunger pangs. If you like chocolate, try sniffing the wrapper — three sniffs per nostril, as prescribed in Hirsch's study.

But even if this doesn't satisfy your hunger, it may do you some good. Researchers at England's University of Westminster have shown that pleasant smells can boost the immune system. They measured antibody levels in the saliva of thirty-six subjects after they had sniffed either melted chocolate or rotten pork. Levels rose significantly with the chocolate and fell significantly with the pork. My guess is that the odor of rotten pork also ruined the subjects' appetites.

13. How does a potato clock work?

First of all, what is a potato clock? Well, it's a clock apparently powered by a potato. You can get one in a kit. All you have to do is insert the two metal electrodes that are attached to the clock into a potato and then stand back and watch it start to tick.

Have you ever bitten into a piece of aluminum foil? If you have, and you had fillings in your teeth, then you'll recall that it was a very unpleasant experience. What you really did was create a battery by joining two dissimilar metals through an electrolyte. An electrolyte is a fluid through which an electric current can flow. When the silver in your fillings and the aluminum were connected by saliva, electrons flowed from the silver to the aluminum, generating an electric current. Ouch!

But this bit of chemistry has practical applications as well. It can be used to make a battery, such as the one required for the potato clock. The clock is based on the idea that electrons will flow from zinc to copper when the two metals are connected through an electrolyte — in this case, the potato. If you want to get technical, though, the potato isn't really powering anything. It's merely providing the means for electrons to flow from one metal to another. The power comes from the reaction between copper and zinc.

Now that you've learned how to make a battery, you also understand why you shouldn't let your silver utensils touch aluminum pots in the dishwasher. Electrons will flow from the silver to the aluminum, destroying the surface of the silver.

14. What are smelling salts?

Victorian ladies fainted regularly. Characterized as the weaker sex, they were conditioned to fulfil the role. A little shock, such as that afforded by a perusal of the novel *Lady Chatterley's Lover*, would bring on a swoon. Someone would rush to find the smelling salts, open the little bottle, and wave it under the victim's nose. Recovery — at least as it's depicted in the movies — was almost immediate. The remarkable substance that resuscitated swooning ladies so effectively was ammonia vapor.

The smelling salts bottle contained a mixture of ammonium bicarbonate and ammonium carbamate, which together are known as ammonium carbonate. These chemicals decompose on exposure to air and release ammonia gas as well as carbon dioxide. The smell of ammonia can quickly bring someone out of a faint.

Compounds that can liberate ammonia have a fascinating history. Many centuries ago, desert nomads noticed that, when they burned dried camel dung, the ancient fuel of desert peoples, a white substance sublimed from the soot that formed. They called it "sal ammoniac" after Ammon, the patron god of the Egyptian city of Thebes, and it became prized as smelling salts. By the Middle Ages, people had discovered that they could isolate the stuff from any burned animal matter — dried vipers were a favorite. The unusual source of the substance undoubtedly lent it a mystical aura and credibility as a medicinal substance.

Today, we understand that smelling salts have no real medicinal value, and we no longer use them. But ammonium carbonate has not disappeared. Bakers still employ it as a leavening agent. Since it releases gas as it's heated, they use it to make cookies and crackers more porous. So, if someone near you swoons and there's no dried camel dung or dried viper available, try crumbling a cracker under his or her nose.

15. Why do some people pee red after eating beets?

Beets are pretty interesting — and not just because about 15 percent of people who eat them produce colorful wastes. For one thing, they are an excellent source of sugar.

Their high sugar content was first noticed in 1747 by Andreas Marggraf, a German apothecary who had been using beets as a laxative. He was struck by their sweet taste. One of his students, Franz Karl Achard, decided to grow several beet varieties to see which produced the most sugar. By 1802, a pilot sugar beet refinery had been built in Prussia.

The real stimulus for producing sugar from beets arose during the Napoleonic Wars. The British had blockaded French ports, cutting off sugar shipments from the Caribbean, so Napoleon ordered his minions to find a way to make sugar beet refineries efficient. One Jules Paul Benjamin Delessert succeeded in finding an effective way to extract sugar from beets, and he was rewarded with the Cross of the Legion of Honor. Soon, forty beet refineries were supplying France with sugar, and that number expanded to 250 by the mid-1920s. Sugar went from being a luxury item prescribed as a sedative for insomniacs to being a widely available commodity. To this day, beets are a major source of sugar in Europe.

Now, back to beets and urine. The naturally occurring pigment in beets, called betacyanin, is a deep red color. It can give you quite a start if it shows up in your urine, but this doesn't happen to everyone. The Technicolor effect is only seen in people who do not have the ability to metabolize betacyanin, and this is a genetic trait. There are no health consequences to this phenomenon.

Urine color can change for other reasons as well. Blackberries — believe it or not — can turn acidic urine red. That's because these berries contain a natural indicator that is black in alkaline solution but red in acidic. And rhubarb can turn alkaline urine red. Of course, you should always see your doctor if your urine is red, because it may be due to blood. But it may just be due to something as simple as what you've been eating.

16. Barbie dolls have been accused of affecting children's health. How?

Poor Barbie. Children love her, but grown-ups just won't stop picking on her. She's been accused of all kinds of crimes. Researchers at University Central Hospital in Helsinki took Barbie's measurements and concluded that her thighs, hips, and stomach were too small. As a real live woman, she'd lack the amount of body fat — 20 percent — necessary to have regular periods. Little girls who tried to emulate Barbie could become candidates for anorexia nervosa, the researchers claimed. Give me a break.

Barbie was actually the toy industry's first "full figure" doll when she appeared in 1958. She was named after Barbie Handler, the daughter of Ruth Handler, who founded the Mattel toy company in 1945. Ken, named after Handler's son, was created to keep Barbie company in 1961. Both dolls became very popular, and out of their success arose some truly bizarre ideas.

None was more bizarre than that put forward by a woman called Barbara, of San Anselmo, California (where else), publisher of the *Barbie Channeling Newsletter*. "I channel Barbie," she insisted — "the archtypical feminine plastic essence who embodies that stereotypical wisdom of the 60s and 70s. Since childhood I have been gifted with an intensely personal, growth oriented relationship with Barbie, the polyethylene essence who is 700 million teaching essences. Her influence has transformed and guided many of my peers through prepuberty to fully realized maturity. Her truths are too important to be prepackaged. My sincere hope is to let the voice of Barbie, my inner nametwin, come through. Barbie's messages are offered in love."

I won't dwell on the implications of all this, but I will mention that Barbara didn't get her chemistry right. The original Barbie dolls were made not of polyethylene but of polyvinyl chloride, or PVC. When this plastic was made commercially available in 1942, it became the basis of a whole slew of vinyl products. The problem with PVC is that it is extremely brittle. In order to give it flexibility, manufacturers mix it with substances that can account for as much as 70 percent of the product's total weight. At one point, plasticizers such as dibutyl phthalate were used to separate the long polymer chains of PVC, allowing them to slide over one another, making the plastic pliable.

Unfortunately, over time, the plasticizers can leach out and form a sticky layer on the plastic. This is worrisome, because such substances have been linked with estrogenic effects, which (with a stretch of the imagination) may have an impact on children who handle old dolls or — even worse — put them in their mouths. It's also a problem for doll collectors and museum curators, since the leaching causes the plastic to degrade and crack. As PVC breaks down, it releases hydrochloric acid, which speeds degradation.

What's the remedy? Light and heat are poisonous to old PVC dolls, so refrigeration is one solution. Another approach is to eliminate the hydrochloric acid as soon as it forms by placing containers filled with a molecular sieve, such as type 4A zeolite, in the dolls' display case. Zeolites are calcium and aluminum silicates that have been heated to eliminate water. This process opens up spaces in the silicate structure into which molecules such as hydrogen chloride can pass and become entrapped.

The leaching problem was not apparent in 1976, the year that Barbie dolls were sealed in a time capsule commemorating America's bicentenary. In 2076, tricentenary celebrants will open the capsule and marvel at the twentieth-century memorabilia it contains. Having been stored in a dark place, the dolls will likely be in decent shape, but they will have lost some weight due to plasticizer loss. Too bad. One thing Barbie cannot afford to do is lose more weight.

17. What would happen to you if you molested a bombardier beetle?

You would be sprayed with a hot solution containing irritant chemicals known as benzoquinones. In all likelihood, this would be a memorable, if unhappy, experience.

Given their ability to discharge these chemical bombs when threatened, bombardier beetles are aptly named. Beetles differ from other insects in that they cannot fly instantly. They store their wings under covers, which they must retract before they can take to the air. Sort of like Clark Kent having to shed his everyday clothes before becoming Superman. Since beetles cannot fly to safety the second they are attacked, they have evolved emergency defenses to deploy while they prepare to flee.

Scientists have studied the African bombardier beetle extensively in order to understand its remarkable defense system. When attacked by predators, mostly ants, the beetle unleashes bursts of hot chemicals with audible detonations. The spray originates from a turret-like appendage under its abdomen, which the beetle maneuvers to achieve remarkable target accuracy. But the truly amazing thing is the chemistry of the spray.

The beetle concocts the irritant chemical just prior to launch by mixing the contents of two separate glands. One contains hydrogen peroxide and hydroquinone, and the other harbors a blend of enzymes, known as catalases and peroxidases, that reacts with hydrogen peroxide to create oxygen gas and water.

When the contents of the two glands mix, oxygen forms and reacts with hydroquinone to convert it to benzoquinone. This reaction is so highly exothermic that the chemical mixture can reach a temperature of 100°C (212°F). Pressure due to the buildup of oxygen then causes the hot mixture of water and benzoquinone to be expelled with a "pop," much to the woe of any attacking ants.

A bombardier beetle can launch up to twenty of these chemical bombs before running out of ammunition. But by that time, it will have succeeded in unfurling its wings, and it's ready to leave its attackers wallowing in its toxic wake. While bombardier beetles can escape from ants using such tactics, they have not been as successful at evading those who argue that the existence of these arthropods proves the theory of creation. Why would separate glands have evolved, the creationists ask, when there is no clear evolutionary advantage until their contents are mixed? The beetle must have been created as is, ready to fight off predators.

Evolutionists don't buy the argument. They maintain that the beetle is an excellent example of survival of the fittest. Random mutations over many years resulted in the protective mechanism that increased their chances for survival — the essence of evolution.

But nobody can contest the fact that the bombardier beetle is in possession of an impressive chemical weapon. So, if you encounter one, leave it be.

18. When are cockroaches treated with the local anesthetic lidocaine?

_____

When they crawl into a person's ear.

Nobody likes to think about bugs crawling into a bodily orifice, but it happens. Cockroaches seem to prefer the ear, perhaps attracted by the gustatory delights of earwax. Unfortunately, the roach's anatomy is such that it affords easy entry into the ear canal but makes for a difficult exit. And the harder the creature struggles to retreat after its earwax meal, the more stuck it gets. Most people don't regard their ears as wildlife preserves, so they seek immediate help to remove the intruder, especially when the cockroach's struggle for freedom results in an earache or tinnitus. While cockroach-in-the-ear is not an everyday phenomenon, it happens frequently enough to have been addressed in the medical literature.

Most physicians confronted with a stuck roach attempt to flush it from the patient's ear with a liquid. The patient may be screaming, "I don't care how you do it — just get that sucker out of my ear!" but the physician must consider the options. Water is not very effective, because a drowning roach will try to hang on to anything it can grab hold of, including an eardrum. Mineral oil has proven more successful. It, too, drowns the roach, but its lubricating effect makes it hard for the roach to hang on. Lidocaine, a local anesthetic, is another choice; the idea here is that an anesthetized roach will not struggle, making it easier for the physician to perform the extraction with tweezers.

Which method is best? Judging by a case report in the _New England Journal of Medicine_, it's the lidocaine method. In 1985, a patient showed up in an emergency room with a unique problem: a cockroach in each ear! The emergency-room physicians immediately recognized a wonderful opportunity

for substantive research and began a double-ear study. They poured mineral oil into one auditory canal and a 2-percent lidocaine solution into the other. The oil did its job, allowing for an easy roach extraction, but the lidocaine performed in a superior fashion. In the report, the cockroach was described as exiting the ear canal "at a convulsive rate of speed." The traumatized roach tried to scuttle off, but a "fleet-footed intern" applied "the simple crush method," and that roach was history. Lidocaine was quickly established as the prime treatment for cockroach-in-the-ear.

Five years later, a Tokyo physician faced with the classic cockroach problem knew just what to do. He flooded his patient's ear with lidocaine, but no roach emerged. Peering into the ear canal, he saw an immobilized bug, which he promptly removed. What was the difference? The Japanese physician had used a 4-percent solution of lidocaine. So, when it comes to driving cockroaches from the ear, the concentration of lidocaine matters.

19. Why are earwigs called earwigs?

The name comes from a European superstition that the insect enters the ear of a sleeping person and bores into the brain. I'm sure you've heard the tale:

An earwig crawls into a lady's ear while she naps on a beach. She doesn't realize that anything is wrong until she starts to experience terrible pain. An x-ray analysis reveals that the bug is burrowing through her brain, and doctors tell the poor woman that the earwig will eventually emerge from her other ear. And that is just what happens. The bug comes out, and the pain disappears. Life goes back to normal for the lady until the pain returns. Her doctors take another x-ray and deliver some devastating news. The earwig was a pregnant female; she laid her eggs inside the patient's head, and now freshly hatched earwigs are devouring her brain! An urban legend, of course. Earwigs may occasionally crawl into ears, but they most assuredly do not bore into brains.

Sometimes, however, truth is stranger than fiction. A Greek physician had a visit from a patient who developed a strange sensation in her ear while on a motorcycle ride. The physician was shocked to see, inside her ear, a spider's web with a spider ensconced in it, apparently comfortable in its warm surroundings. Recognizing that this was an epic moment, he ran for his video camera and managed to record the arachnid's hasty departure from its new home.

Don't think that such things happen only in Greece. A similar incident occurred in Nova Scotia, Canada. A lady complained to her doctor of a buzzing in her ear — she feared that a fly had somehow flown in. Once again, a stunned physician came face to face with a spider. The Nova Scotia doctor initiated the usual treatment for bugs in the ear, squirting a little water into the ear canal. But the spider didn't take kindly to this at all. It jumped out of the ear and ran down the patient's face, causing her to hyperventilate and run around the examination room yelling, "Oh, my God! Oh, my God!" Could have been worse. Could have been a pregnant earwig.

20. Aristotle used goat urine, and Hippocrates recommended pigeon droppings. For what?

Both of these substances were reputed to cure baldness.

Most men don't find baldness an appealing trait, despite the stories of bald men's sexual prowess (stories likely initiated by bald men). Throughout history, men with bald pates have tried anything and everything to stimulate hair growth. The ancient Egyptians applied rancid crocodile or hippo fat, rationalizing that, if it smelled bad, it must be doing some good. It wasn't. In order to coax Julius Caesar's dormant hair follicles into action, Cleopatra experimented with a goo made of ground horse teeth and deer marrow. When this didn't work, she traded him in for Marc Antony. Bald Victorians brushed cold tea, followed by citrus juice, onto their scalps. In farming areas, chicken droppings were used, and baldies even persuaded cows to lick their heads. Electric combs, suction caps, and paint thinner were all tested as baldness remedies.

There is no end to it. Infomercials push shampoos with special emulsifiers that clean follicles, as if baldness were due to plugged follicles. Others promote what amounts to a spray paint that covers bald spots.

The truth is that only Rogaine (minoxidil) rubbed into the scalp, and Propecia (finasteride) taken orally, have shown any hair-growing effects, but these effects are not very impressive. The Bald-Headed Men of America — headquartered, appropriately, in Morehead City, North Carolina — was established by a man who was refused a job because he was bald. Members of this organization take a unique view of their condition. "If you want to waste your hormones growing hair," they say, "then go ahead."

In fact, this notion is wrong, because a cause of baldness is high levels of the hormone dihydrotestosterone. The Bald-Headed Men of America are on firmer footing when they voice the slogan "No rugs or drugs."

21. How do cap pistols produce their bangs?

"Bang! Bang! I got you!" How many millions of kids have yelled that while playing cowboys and Indians or cops and robbers or whatever other violent game was in vogue? The guns, of course, were not real, but the bangs were.

And they came fast and furious, limited only by the length of the rolled-up red strip speckled with black dots that was loaded into the cap pistol's heart. There is actually a lot of technology behind the bang. The cap's major ingredients are red phosphorus, sand, potassium chlorate, and manganese dioxide. When the pistol's hammer strikes the cap, the grains of sand rub against each other and generate enough heat to ignite the phosphorus. Heat also causes the potassium chlorate to decompose and release oxygen, which then supports the combustion of the phosphorus. As the phosphorus burns, it quickly heats up the surrounding air, which expands and causes a sound wave. Bang! And what about the manganese dioxide? It's a catalyst that makes the reactions involved proceed more rapidly.

22. What is the meaning of the term patent medicine?

———

The most important aspect of our lives is our health. These days, when something goes wrong with our bodies, many of us seek medication to remedy the problem. Modern science has given us a host of effective drugs to turn to, but the practice of ingesting foreign substances to treat illness is ancient.

Hippocrates recommended willow bark for pain, which made sense, and pigeon droppings for baldness, which did not. Since his day, a staggering number of remedies have been identified and concocted. But it was only during the twentieth century that governments began to introduce regulations to ensure that drug manufacturers' claims were legitimate and that their products would do more good than harm to the consumer. These regulations were prompted by the bewildering array of uncontrolled medicines that had flooded the market in the late 1800s, the era of the patent medicines.

Why "patent" medicines? Because in mid-eighteenth-century England, some producers of medical preparations applied for, and obtained, royal patents for their products. The patents protected the owners' rights to the products and lent those products a certain prestige. Manufacturers were under no obligation to demonstrate safety or efficacy. Later on, the term *patent medicine* was applied to any mass-market product with unregulated ingredients, promoted through unregulated advertising, that was used to treat common human ailments.

Lydia Pinkham's Vegetable Compound was the most successful of all the patent medicines. Mrs. Pinkham became interested in home remedies after several of her family members died. She turned to spiritualism and to chemistry. Convinced that God had provided vegetables and herbs to cure disease, she blended these natural substances with a good measure of alcohol and provided satisfaction to a lot of women. They likely derived most of their satisfaction from the alcohol, but Lydia's original preparation did include black cohosh, which we now know can alleviate some of the hardships of menopause. When Lydia's son put his mother's picture on the vegetable compound's box, modern advertising was born.

Although patent-medicine vendors tended to promise far more than they could deliver, their products often included active ingredients, such as opium or alcohol. One such product was Paregoric, an alcohol solution flavored with camphor and aniseed; its name derives from the Greek word for "soothing." Paregoric was used to treat cough and diarrhea, and it did so pretty effectively. Some patent medications contained datura stramonium and belladonna extracts. The active ingredient in these is atropine, which would have had some effect against asthma. Quinine was useful in treating fever, and some patent salves contained phenol, a disinfectant.

But the vast majority of patent medicines were nonsensical nostrums. Stomach bitters contained an unspecified mix of barks, roots, and herbs. Dr. Chase's Syrup of Linseed and Turpentine bore a label that didn't even explain what the syrup should be used for, and all that Bodi-Tone claimed to be was a "tonic for sick bodies." Some remedies had very odd ingredients — for example, Four Chlorides Compressed Tablets contained arsenic.

The patent-medicine era was fascinating indeed. What folly, we think in retrospect, to swallow untested, unregulated products purely on the basis of imaginative advertising. But do you know what? In a sense, the patent-medicine era lives on. The Internet is filled with pitches for untested nostrums that recall the glory days of patent medicines.

New York Stress Tabs is a case in point. According to the label, this product can "manage daily stresses related to sleep, work, relationships, travel, hangover, overindulgence, and PMS."

What magical ingredients does it contain? Aconite, which is the poison that Romeo and Juliet imbibed, as well as strychnine. Delightful! But don't worry, these ingredients are present in "homeopathic doses," which means that their concentration is roughly zero.

23. How does a radiometer work?

You've seen this little device. It looks like a lightbulb with a weather vane inside it. Hold it up to a light source, and the weather vane spins.

The radiometer can be found in any science shop, and it's been amusing people of all ages for over a hundred years. It was invented by William Crookes, the scientist who was better known for inventing the cathode-ray tube, which eventually led to the development of x-ray machines.

The effect relies upon the colors of the vane. One side is white, and the other is black. When exposed to light, the black side absorbs more of it than the white, and some of this light energy is converted to heat. When you wear white clothing in the summertime, you're responding to the same principle — your white T-shirt reflects the light, but you'd cook if you wore a black one.

Now, when the air molecules (essentially oxygen and nitrogen) inside the bulb strike the warmer, black surface, they themselves absorb some of the heat energy and start to move faster. They kick off the surface at a faster speed than the speed they arrived at. The molecules that bounce off the white surface do so with considerably less energy. Since, as Newton told us, for every action there is an equal and opposite reaction, the vane will start to turn as if it's being pushed from the dark side. The more light there is, the faster the vane will spin. It might even make your head spin. Maybe that's what it did to William Crookes.

Although he was a noted scientist, Crookes became involved in spiritualism. He was fascinated by the celebrated Fox sisters, who, in 1848, at the ages of twelve and fifteen, claimed to have heard strange "rappings" in their bedroom. The Fox family called in witnesses. These people did hear the rappings, which seemed to emanate from the girls themselves, but no one could determine their exact origin. They concluded that the phenomenon had to be some sort of message from the spirit world. The girls were sent on a highly publicized tour. Sir William Crookes saw them, and he, too, speculated that the rappings were a message from the other side. It seems that he, along with many others, failed to realize that the girls were merely cracking their toes!

Sir William was apparently hoodwinked by one Annie Eva Fay, as well. With her husband, Fay put on an intriguing stage performance. Volunteers from the audience would tie her securely to a chair, and she would appear to fall into a trance. A curtain was drawn around her. Behind the curtain, although her hands were tied, Fay busily played guitar, snipped paper dolls with scissors, and hammered nails into a piece of wood. Fay allowed Crookes to put her to the test. To keep her from slipping her bonds, he used an electrical circuit that had two handles. He ordered Fay to grip both handles for the duration of the test; an ammeter would detect any interruption in the current. She performed to Crookes's satisfaction, and he declared her genuine.

Later in her career, Annie Eva Fay met the great Harry Houdini, and she told him how she had tricked Crookes by securing one handle of the circuit beneath her knee and using her free hand to perform her various tasks. Years after that, one of Crookes's galvanometers was discovered in the London Science Museum. Tests showed that a subject could insert both electrodes in his socks, freeing both hands, without interrupting the current.

Was Crookes really taken in, or, as some have suggested, did he simply delight in playing around with young women under his wife's nose? We'll never know. But one wonders how someone clever enough to invent the radiometer could be so easily fooled.

24. Why is Scotch tape called Scotch tape?

The Scots did not invent Scotch tape. But their reputation for thriftiness did inspire the name of this amazing product.

Two-tone cars were the in thing way back in the 1920s, and car manufacturers were faced with the problem of how to cut clean, crisp lines between the colors. Before spraying on the paint, they would mask one side of the line with newspaper to create a sharp, straight edge. This worked well, except for the fact that it was hard to remove the glued-on newspaper after the job was done.

At the time, 3M was selling sandpaper to car manufacturers, and the company's salespeople heard about the problems the manufacturers were encountering in their paint shops. A great potential market beckoned for a sticky tape that peeled off easily. A 3M chemist named Richard G. Drew rose to the challenge of developing such a product. Rubber cement, he knew, had the necessary properties: it was sticky, yet one could peel it off a surface fairly easily. Drew managed to coat one side of a paper strip with the material, and he was satisfied to see that, by applying a little pressure, he could make the paper adhere to a surface; with equal ease, he could remove it. He figured that 3M could produce the tape cheaply, especially if they applied the glue only to the edges of the paper strip — no need to waste glue.

The car painters thought the newfangled tape was a great idea — that is, until they started to use it. There wasn't enough glue to hold the tape firmly in place, and the 3M tape salesmen were unceremoniously told to take the tape back to their bosses and tell them to be more generous with the adhesive next time. The 3M company quickly fixed the problem, but the stigma of the failed first attempt lingered. The car painters took to calling the improved product Scotch tape, and 3M was stuck with the name. Wisely, they decided that, if you can't beat 'em, join 'em and went on to develop a whole line of Scotch tapes. When they got around to applying glue to clear cellophane, see-through Scotch tape was born.

Today, over four hundred different varieties of pressure-sensitive tape are available. Manufacturers employ various glues, but most of these fall into the acrylic family of polymers. They are not designed to be removed as easily as masking tape, and they adhere strongly because they produce numerous microscopic suction cups when pressed on a surface.

Indeed, some tape adhesives are so strong that they create a problem. One example is the adhesive used on bandages, and that's why removing a bandage strip can be a painful experience. But there may be a solution in the offing. Researchers in England have developed a new pressure-sensitive bandage with an adhesive that can be deactivated with light. When you peel off its opaque backing, light initiates a reaction in which side groups on the polymeric acrylic molecules link together, destroying the adhesion. A camera flash removes the bandage — in a flash!

25. What is the chasteberry, and why is it so named?

Technically, this shrub, which grows around the Mediterranean, is called *Vitex agnus-castus*. But the plant's common name — the chasteberry — is more intriguing. Folklore has it that the berries suppress the libido in both men and women. Indeed, monks once chewed them to diminish temptation.

Today, the chasteberry is used for a different purpose. While the berry contains no hormones, it can influence blood levels of progesterone. And what is the good of that? An increase in progesterone may be just the thing for women who suffer from premenstrual syndrome, or PMS. Scientists have linked the irritability, depression, and bloating that often plague women during the two weeks prior to menstruation with lowered levels of progesterone. These symptoms may respond to treatment with the chasteberry. Some studies have even shown benefits for women who suffer from menopausal symptoms or from fibrocystic breasts. According to the best evidence, about four hundred milligrams of a powdered extract standardized to 0.5 percent agnuside, the active ingredient, is the appropriate daily dose.

Those on hormone-replacement therapy or the birth-control pill should not consume chasteberry, because combination effects are unknown. But women who have difficulty achieving pregnancy due to irregular ovulation may benefit from chasteberry. Their difficulties may be due to too much prolactin and too little progesterone, both of which can be regulated with chasteberry. They must, however, discontinue the therapy as soon as they conceive, since the effects of chasteberry compounds on the developing fetus are unknown.

But will those consuming the chasteberry feel motivated to do what's necessary to conceive? Sure. Modern science has shown that the monks were wrong. Chasteberry does not interfere with the sex drive.

26. At one time in parts of Scandinavia and Germany, butchers were prohibited from hanging slaughtered hares and rabbits in their shop windows. Why?

————

Long ago, many people believed that the baby of a pregnant woman who looked upon a slaughtered animal would be born with a cleft palate.

It is difficult to know how such superstitions get started, but most can be traced to our innate desire to find causes for the calamities that befall us. When there is no ready scientific explanation, we invent one. That's why the Black Plague was blamed on witches and Jews, the waxy pallor of tuberculosis victims was blamed on vampires, and migraines were blamed on the evil eye.

People need to identify cause-and-effect relationships, even when they don't exist. Today, instead of blaming our ailments on witches and evil eyes, we blame them on chemicals. The person afflicted with a headache may identify the culprit as a fabric-softener sheet, because she read all about it on the Internet. Another may blame cleaning agents for his health problems — or pesticides or cookware or microwave ovens or antiperspirants or cell phones. Perhaps, however, the real cause of their suffering is the bizarre medical experiments they underwent at the hands of aliens who kidnapped them and wiped their memories clean.

There is no more evidence for most of these connections than there is for the one between cleft palates and butchered animals. We still don't really know why the upper lip and palate fail to close completely in about one in seven hundred unborn babies. Certainly, there is a genetic aspect to this, but environmental factors come into play as well. Tests have implicated alcohol intake during pregnancy, as well as the use of the cough suppressant dextrometorphan. Seizure medications, the acne drug isotretinoin, excessive vitamin A consumption, certain herbal remedies, pesticides, folic acid deficiency, and smoking during pregnancy have all been accused of playing a role in cleft palate.

The only effective treatment for cleft palate is surgery. The condition is also called harelip, because those who are afflicted with it have upper lips that resemble a hare's. This is probably what gave rise to the superstition about pregnant women who gaze upon slaughtered animals, and it could explain why the butchers of yore didn't display their slaughtered hares.

27. How was O. J. Simpson's defense team helped by their client's taste for tacos?

———

Remember what a stir the stained Ford Bronco caused during the O. J. Simpson murder investigation? Were the stains blood, or — as the defense asserted — were they taco sauce?

Criminal investigators presented a strong case for blood, basing their theory on phenolphthalein. Here's the story. Phenolphthalein is a classic acid-base indicator that is colorless in acid and a bright pink in base. But when it is heated with zinc, it converts to a compound (reduced phenolphthalein) that does not produce a pink color in base. This derivative can be converted back to phenolphthalein (oxidized) with sufficient amounts of molecular oxygen, which can be generated from hydrogen peroxide in the presence of a catalyst.

As it turns out, blood has just such a catalyst. Hemoglobin, found in all red blood cells, is very adept at hastening the decomposition of hydrogen peroxide into oxygen. This is a very important function, because hydrogen peroxide, a by-product of metabolism, is quite capable of damaging tissue.

So, crime-scene investigators treat suspicious stains with reduced phenolphthalein and hydrogen peroxide. If the stain is indeed blood, the hydrogen peroxide releases oxygen, which oxidizes the reduced phenolphthalein to phenolphthalein. This is then treated with base, and if the characteristic pink color appears — bingo! And did the pink color materialize when investigators went to work on the stains in O. J.'s Bronco? Sure did. But the defense argued — pretty effectively, as it turned out — that certain fruits and vegetables also have enzymes capable of liberating oxygen from hydrogen peroxide. Simpson's lawyers contended that taco sauce would have yielded the same test result as blood. Actually, plant enzymes are readily broken down by boiling, but the required blood enzymes are not. The investigators could have turned up the heat on O. J. by applying heat to the sample, but they simply tested for cold blood.

Phenolphthalein did not solve the O. J. case, but it could have prevented a costly fiasco that occurred on a New Jersey highway. In the course of an accident, a car's air bag was deployed, filling the vehicle with a fine white powder. The rescuers, fearing that a toxic substance had been released, cut the driver's clothes off his body and cordoned off the area. They even set up an emergency shower for anyone who came into contact with the powder. Twenty-two people were admitted to hospital for observation.

Why all the concern? Because caustic sodium hydroxide (lye) can form as a by-product of the chemical reaction that inflates an air bag, and officials at the scene thought that the white powder released was sodium hydroxide. It was actually talcum powder, which air bag manufacturers use as a lubricant. The investigators could have determined that sodium hydroxide was not present at the scene of the accident if they'd performed the simple phenolphthalein test.

28. What is meant by "water intoxication?"

The toxicologist's refrain is "Only the dose makes the poison." The most innocuous substances can be deadly in high doses — including water. Believe it or not, there is such a thing as water poisoning. And it can be lethal.

If body fluids become too diluted, then sodium concentrations in the blood drop dramatically. Sodium is an essential mineral in the body, and it plays a critical role in the transmission of nerve impulses in the brain and the muscles. Our bodies regulate the concentration of sodium in our blood by moving water in and out of the blood. If you have a high sodium concentration, then water moves from your cells into your blood, increasing your blood pressure. As water moves from your brain cells into your blood, your brain actually shrinks, and you may experience confusion and seizures. If your sodium concentration is low, then the reverse happens:

water moves out of your blood and into your cells. This can be disastrous to your brain, since, as your cells absorb water, your brain swells. This, in turn, can lead to lethargy, loss of consciousness, and even death.

This isn't merely a theory — excessive water consumption has actually killed. In a truly disturbing case, the parents of a four year old were charged with child abuse and murder. The Utah couple felt that their newly adopted daughter wasn't bonding with them properly. Acting on what they claim was the advice of a misguided alternative therapist, they tied their daughter's hands behind her back and forced her to drink huge quantities of water. The idea was that such "therapy" would transport the girl back to infancy, when she consumed nothing but liquids; in this infantile state, she could be "reprogrammed."

The couple and their lawyers built their defense on this version of events, accusing the therapist of malpractice. But the prosecution insisted that no one ever recommended such hydrotherapy to the couple and that the adoptive parents were in fact punishing the child for stealing some Kool-Aid from a sibling. We don't know what really happened. Was this an instance of alternative therapy gone awry or a case of child abuse and murder? One thing alone is clear: the unfortunate child died of water intoxication.

29. Sodium sulfite is not a cleaning agent, yet it is commonly added to detergents. Why?

To protect washing machines and dishwashers from corrosion. Iron reacts with oxygen to form ferric oxide, which is better known as rust. This reaction proceeds more readily at high temperatures, such as those found in washing machines.

Where does the oxygen come from? It is dissolved in water. Water surfaces come in contact with air, and some of the oxygen dissolves. Oxygen also is a by-product of photosynthesis, which occurs as aquatic plants grow. The amount of oxygen that dissolves depends upon temperature (less dissolves as the temperature increases), pressure (less dissolves at higher altitudes), and the amount of other substances already dissolved in the water (fresh water holds more oxygen than salt water). Oxygen in water is good for fish, because they fulfil their oxygen needs by extracting the gas as water passes through their gills, but it isn't so good for metals in washing machines, because they corrode.

Sodium sulfite is an oxygen scavenger. It reacts with oxygen to form sodium sulfate, effectively lowering the dissolved oxygen content and thereby protecting the insides of washing machines from rusting. Such corrosion may be an annoyance at home, but in industry it's a huge problem. Sodium sulfite is commonly used to keep high-pressure industrial boilers from rusting, but the substance can break down in boilers to form sulfur dioxide along with hydrogen sulfide, which not only smells like rotten eggs but is itself corrosive.

Alternatives are available. Hydrazine, N_2H_4, can be added to boiler water to virtually eliminate dissolved oxygen. It reacts with oxygen to yield nitrogen gas and water, neither of which presents a problem. But hydrazine is pretty nasty stuff to work with, so scientists have sought out less toxic compounds. An interesting alternative is sodium erythorbate, which reacts with oxygen to produce lactic and glycolic acids, which themselves react with oxygen to yield carbon dioxide.

You may have encountered sodium erythorbate while eating lunch — it serves double duty in hot dogs. Sodium erythorbate is a preservative, and it allows manufacturers to use less nitrite in their hot dogs. It is also, as we have seen, an oxygen scavenger, so it prevents the formation of the off flavors that are produced as meat components react with oxygen.

Incidentally, there is absolutely no truth to the rumor that sodium erythorbate comes from earthworms. The rumor probably got started when someone noticed the similarity in pronunciation between "eryth" and "earth" and jumped to the wrong conclusion.

30. Fill in the blank in this curious Malaysian expression: "When the ___ are down, the sarongs are up."

———

The answer is durians. And what are durians? One of the most interesting fruits in existence.

Durians grow mostly in Southeast Asia, where they are referred to as "the king of fruit." The only problem is the king needs a bath. While people describe the fruit's taste in exalted terms, they admit that it smells like a public toilet with a backed-up drain.

Durian is an Indonesian word that derives from *duri*, which translates as "thorn." Indeed, the durian is a thorny fruit in more ways than one. It is about the size of a large cantaloupe, and it's shaped like a football. Its hard shell is covered with sharp spikes. Let's just say that, when a durian falls from a tree, you don't want to be standing under it. But eating the fruit is a different matter entirely.

It seems that the flavor of a ripe, well-chosen durian is exquisite. The choosing, though, should be done by an expert, a durian seller known as a *tukang durian*. The *tukang durian*, always a man, conducts business from a roadside stall. The taste of the *tukang durian*'s wares is often described as that of exotic raspberries — albeit exotic raspberries consumed in a public toilet. Durians smell so bad that they are banned from airplanes, hotels, and public transport. Devotees are not turned off by the smell; in fact, they claim to be turned on by the taste. Legend has it that the durian has aphrodisiac properties, hence the expression "When the durians are down, the sarongs are up."

San Francisco's Marco Polo Ice Creamery makes durian ice cream. According to the hype, the first taste may not enthrall you, but the durian flavor builds with each subsequent bite. Whether the durian arouses the passions is questionable, but durian lovers sure are passionate about the king of fruit.

31. Why can't you use low-fat margarine for frying?

Margarine is fat. It's just a different kind of fat from butter. Whereas butter contains mostly saturated fats, the fats in margarine are a mix of the saturated and unsaturated varieties. Saturated and unsaturated fats vary slightly in molecular structure.

This variation affects our blood-cholesterol levels. Saturated fats increase cholesterol, while unsaturated fats — in moderation — do not. So, if margarine is just fat, how do manufacturers create a low-fat version? Easy. They just mix the fat with water. In a given volume of margarine, some of the fat gets replaced by water, and the margarine can be termed "low fat" or "reduced fat."

Everyone knows that fat and water don't mix, but, when we shake these immiscible components together well, tiny water droplets get suspended in the fat. We can then add emulsifiers to prevent the water from separating. Now for the problem. Water boils at 100°C (212°F), but fat does not — it can be heated to a much higher temperature. Picture what happens. The tiny water droplets convert to steam, but they are trapped inside the matrix of fat. As the temperature increases, the steam pressure builds. Then the water-vapor bubbles try to expand in volume, but they can't because the fat is weighing them down. Eventually, the pressure becomes so great that the steam-filled bubbles explode. You don't want to be standing nearby when that happens, because hot fat will splatter all over the place.

This explains why bacon spits when you fry it and why you should never douse an oil fire with water. Unless, of course, you want to spread it.

32. Why is it customary for a gentleman to remain on the curb side of the sidewalk when walking with a lady?

It's all a matter of physics. And chamber pots. Before the advent of plumbing, people emptied their chamber pots from the upper-story windows of their houses into the gutters that lined the sidewalks below. The trajectory of the contents made strolling hazardous for anyone near the gutter — that is, on the curb side of the sidewalk. Gallant gentlemen took to walking on the outside to prevent catastrophe from befalling their ladies.

Today, we are revolted at the thought of disposing of waste in this fashion, but cleanliness didn't really become a virtue until modern times. When Queen Victoria ascended to the throne in 1837, Buckingham Palace had no baths at all. Perfumes were very popular, since they masked objectionable smells. Madame de Pompadour spent vast amounts on perfumes, and Madame du Barry hid scented pads on her person when she seduced Louis XV. But some were not put off by a ripe body aroma — among them Samuel Johnson, the creator of the first English dictionary. When an outspoken lady friend told him that he

smelled, the gamy Johnson took no offense. He did, however, object to her misuse of language. "You smell," he corrected her. "I stink."

33. How does an eraser work, and why does it erase pencil more easily than ink?

———

Erasers have an interesting history. When the Europeans first arrived in South America, they discovered natives making interesting use of tree sap. They were shaping the wet latex into balls and allowing the balls to dry, thereby creating the world's first bouncing balls.

The explorers took some samples home with them, and one found its way into the hands of the brilliant English chemist Joseph Priestley, who found a use for the material. It would neatly rub pencil marks off paper! He called it the "rubber" — in North America, we call it the "eraser."

How does it work? Through the physical action of abrasion. A pencil deposits a thin layer of graphite on paper, a layer that lifts off when gently rubbed with an eraser. The trick is to remove the graphite without damaging the paper, so we do not want an eraser that is too abrasive. Chemists have worked on this problem, and they have improved on the original rubber. For one thing, they have vulcanized the rubber — that is, they have treated it with sulfur. This increases its resiliency and reduces its rigidity. They have also added a fine pumice powder made of about 75 percent silicon dioxide and 25 percent aluminum oxide to the rubber, giving it the correct abrasive quality. This pumice is the stuff they use to make sandpaper, but they grind it much finer for inclusion in erasers.

Why do erasers not work as well on ink? Unlike graphite, ink is absorbed into paper, so rubbing the surface isn't enough. Deeper penetration is needed, which damages the paper.

Why is the traditional eraser pink? These days, most erasers are made of synthetic rubber, which is a polymer. Manufacturers add a chemical called an accelerator to the rubber during the production process to help join the monomers, or small molecules, into a polymer. The original accelerator was pink in color, and today's manufacturers uphold the tradition by using a pink dye.

Just like everything else, erasers have changed with the times. Most of us, when we think of polyvinyl chloride, or PVC, think of vinyl car roofs or the old-fashioned LP record album. But this synthetic polymer can also be formulated into erasers. To make the material soft and pliable, manufacturers add a plasticizer such as dioctyl phthalate to the mix; to improve abrasion, they stir in some calcium carbonate, or chalk.

Vinyl erasers are gentler on paper than rubber erasers, and they have some other admirable characteristics as well. For example, white vinyl erasers are good for cleaning ivory. And people who are allergic to latex need not worry about reacting to a vinyl eraser when rubbing out their mistakes. There is, however, some cause for concern about the environmental consequences of phthalates — they have been accused of having endocrine-disruptive effects. There's no free lunch. But I won't rub that in.

34. Why do lobster and shrimp change to a reddish-orange color from a grayish-blue color when we cook them?

Shrimp and lobster have something in common with carrots: they contain a good dose of the yellow-orange compounds called carotenoids.

These compounds are quite widespread in nature, but the first one that scientists isolated was beta-carotene from carrots, so they bestowed that name on the whole family. Carotenoids are also responsible for the color of orange juice, red peppers, watermelon, tomatoes, egg yolks, apricots, corn, pink grapefruit, pink salmon, and pink flamingos.

Lobster and shrimp dine on plankton that contain carotenoids, and the compounds become concentrated in their shells. There, they are bound up with protein molecules, and the carotenoid-protein complex is dark green. Cooking the shellfish heats the protein and denatures it. In other words, the protein breaks down and disassociates itself from the reddish carotenoid, astaxanthin, which then becomes visible. To a lesser extent, this process is also evident when we cook carrots — they become more orange than they were in their raw state. The effect is not as great as it is with shrimp and lobster, because carrots do not have much protein.

Carotenoids extracted from natural sources are widely used as food dyes. Annatto, a tropical shrub extract, is intensely colored due to the presence of the carotenoid bixin, and it can be used to color Cheddar cheese, margarine, and butter. This dye is completely harmless; in fact, the natural yellow color of butter is due to carotenoids found in grass. When grass dries into hay, these compounds are destroyed. Winter butter, therefore, tends to be white, and producers often add carotenoids to it to increase its market appeal.

Obviously, then, when a lobster is plunged into boiling water, a lot of interesting chemistry occurs. I doubt, however, that the lobster appreciates it.

35. It's the mid-1800s. A large crowd surrounds Hal the Healer's platform. Before he begins his performance, Hal signals to his band to strike up a lively tune. What is he preparing to do, and why does he need music?

Hal was preparing to pull a tooth, and he needed the music to drown out the screams of his patient.

In the nineteenth century, tooth pulling was a common public attraction. Everyday life was pretty dreary, and watching someone else suffer provided a little distraction. Dentists became public entertainers, vying with each other for business. The more flamboyant the performance, the bigger the crowd and the greater the opportunity for drawing in more vic — er, patients. Hal the Healer used music to drown out his patients' muffled screams. Why muffled? Because Hal's advanced technique involved pulling the tooth with his right hand while choking off the patient's windpipe with his left.

Diamond Kit, who performed adorned with jewelry, used a different procedure. He liberally administered a "pain reliever" to his patients, swabbing it over their mouths before the extraction. When Kit yanked a tooth, not a peep was heard. But this had nothing to do with his so-called pain reliever — it had everything to do with the large wad of cotton he stuffed into his victim's mouth as soon as he'd performed the extraction.

Dr. Jean St. Pierre, who performed his dental artistry as part of a traveling medicine show, had a different twist. Not only did he pull in customers with his tooth-pulling performance, but he also offered to treat other members of the show troupe for free. Then the show's owner noticed that his performers were giving substandard performances and seemed to be functioning in a daze. When he asked St. Pierre to extract one of his own molars, he discovered the reason. The doctor gave him some pills that packed a hefty dose of opium. It seems that St. Pierre was generating a little income on the side by peddling opium to the performers who had become hooked after experiencing his "painless dentistry."

36. Why does rubber have elastic properties?

In 1823, a London coach maker named Thomas Hancock took rubber sheets, cut them into strips, fastened the ends together, and invented the rubber band. Of course, he had no idea why rubber behaved like rubber.

Today, we understand that rubber is composed of polymers, large molecules — not large enough to see, of course, but large in relation to other molecules. The long chains of carbon atoms that make up the rubber molecules feature some double bonds; in other words, in certain places, adjacent carbon atoms are held together very strongly, producing angles in the chain. The result is that the molecules have natural kinks in them. Stretching the rubber straightens the molecules, which then tend to recoil to their original shape. This tendency is what we call elasticity.

When you stretch a rubber band, you use energy to uncoil the molecules. Some of this mechanical energy is transformed into heat, warming the band — prove this to yourself by touching it to your lips. When you allow the band to revert to its original shape, the molecules will do the work for you. Some of this is the mechanical contracting, and some is the absorption of heat from the air. The elastic band gets cold.

There is a lot of science to be learned from rubber. Any material that is distorted by a force and resumes its shape when the force is removed is said to be elastic. When we squeeze a material, we can observe elastic compression. When we throw a ball on the floor, it squeezes flat at the bottom and bulges on the sides. Elasticity means a tendency to round into the original shape. Clay, for example, is not elastic at all. Drop a clay ball, and it will not regain its form. Glass and steel, you may be surprised to learn, are elastic, but they don't compress much, so they don't have much bounce.

Today, we have synthetic rubbers with molecules that rebound to their original shape even more efficiently than those of natural rubber — hence the "superball." We also have synthetic rubbers with molecules that absorb energy and do not release it again. Balls made of these substances feel the same as other rubber balls, but, when their molecules are deformed, they do not instantly bounce back to their original form. It's all a matter of molecular structure, and that's the way the ball bounces. Or doesn't.

37. In the 1960s, scientists examined the memory of worms, conducting a series of highly publicized studies that focused on feeding worms a special diet. What was that diet?

————

Other worms. James McConnell of the University of Michigan, who would eventually publish a journal called *Worm Runner's Digest*, got the ball rolling by showing that flatworms could be conditioned in a fascinating way.

McConnell claimed that, if he exposed a worm to light followed by an electric shock, the worm would alter its slithering path. His worms associated the light with the shock and learned to avoid the path that delivered the shock when the light was turned on. McConnell also knew that, if you cut a worm in half, both halves would regenerate, and he wondered which half would retain the memory. To his amazement, he discovered that both did. The researcher had to conclude that memory was somehow stored in molecules throughout the worm's body.

Then McConnell was struck by another idea. Grinding up trained worms, he fed them to untrained worms. The cannibal worms began to avoid the light as well. Apparently, memory had been transferred. This astounding observation stimulated much research, not only in worms but also in fish, mice, and rats. Unfortunately, other researchers could not reproduce McConnell's results, and as yet nobody has isolated memory molecules.

But the worm research did have some interesting spin-offs. Mice that were fed the brains and livers of other mice subjected to stress — researchers had rolled them around in glass jars — learned to associate a light stimulus with electric shock faster than other test animals. It seems that stress hormones transferred by diet enhanced learning.

Maybe someone should examine whether humans who dine on stressed, factory-raised chicken are smarter than people who favor free-range, carefree poultry. Maybe they are, since they are unwilling to pay higher prices for chicken that has no clear superiority. Of course, someone is bound to point out that smart people don't eat animals in the first place. Maybe we should stress broccoli and see what happens.

38. In the mid-1800s, a French industrialist who distilled alcohol from sugar beets ran into a problem when some of his fermentation vats stopped producing alcohol. He asked a scientist for help, and this resulted in one of the most important scientific discoveries ever made. Who was the scientist, and what was the discovery?

———

Louis Pasteur's involvement in the problem led to the discovery of the role bacteria play in disease.

Pasteur, a young chemistry professor at the University of Lille, was saddled with a problem that dogs many contemporary researchers: he lacked funding. Private manufacturers he approached for help were often skeptical about the need for research, but Pasteur won many of them over to his cause with his thrilling lectures.

At one such lecture, Pasteur asked his audience where one could find the young man "whose curiosity will not immediately be awakened when you put into his hands a potato, and when with that potato he may produce sugar, and with that sugar alcohol?" These words struck a chord in one Monsieur Bigo, the industrialist whose beets had stopped producing alcohol, and Bigo begged Pasteur to look into the problem.

At the time, nobody — Pasteur included — understood how sugar ferments into alcohol, so the chemist did the only thing he could think of: he took a sample from a healthy vat and put it under a microscope. A swarm of yellow globules danced before his eye. Not only did they dance, but they also multiplied. These yeasts were alive, and, as Pasteur quickly surmised, they converted sugar to alcohol.

Next, Pasteur examined the sick fermentation vats. Instead of yeast globules, he saw tiny, rod-like things cavorting in the broth, which instead of alcohol was filled with lactic acid. Pasteur managed to transfer some of the rod-like creatures into a clean container, and he went on to demonstrate that, when they are given appropriate nutrients, they multiply. The rods were alive, and they busily converted sugar to lactic acid. These microbes had contaminated the beet-sugar vats, killed off the yeast, and terminated the production of alcohol.

Where had they come from? Pasteur correctly surmised that they were present in the air. If they could lead beet sugar astray, what else could they do? Pasteur soon answered this question by establishing a link between microbes and human disease. And all of this came about because a French industrialist had a problem with alcohol production.

39. In the late seventeenth century, Dr. Nooth's Apparatus was found in almost every well-to-do English household. It was designed to maintain good health and restore the health of those who had lost it. What did the apparatus do?

———

Produce soda water. People had long believed that naturally carbonated waters were healthy. Those suffering from kidney stones, arthritis, and "lack of vigor" flocked to spas to partake of the waters. But individuals who were unable to get to the source were out of luck. That is, until Joseph Priestley came along.

Priestley, best known as the discoverer of oxygen, lived next to a brewery and was intrigued by the bubbles of carbon dioxide he saw rising in the beer. This gave him the idea of carbonating water artificially. Joseph Black had already produced carbon dioxide — or "fixed air," as it was called — by reacting chalk (calcium carbonate) with sulfuric acid. Priestley designed a clever apparatus that linked a glass vessel containing these reagents to a pig's bladder. The bladder was connected to a tube that fed into a water-filled bottle that sat inverted in a basin of water. The gas was generated, and it filled the bladder, which was then squeezed to pump pressurized gas through the water. Using this method, Priestley was able to dissolve enough of the gas to produce an acceptably bubbly beverage. Salts such as sodium carbonate or sodium tartarate could be added to produce mineral water.

John Nooth, a Scottish physician, wondered why the curative properties of Priestley's water had not been extensively investigated. He suspected that it had to do with the fact that the water had an unpleasant, urine-like flavor, so he designed an apparatus made entirely of glass, eliminating the pig's bladder. Priestley did not take the criticism of his water very well. He maintained that neither he, nor anyone to whom he'd served his water, had ever noted a urine smell or flavor. If Nooth had found his water unpalatable, Priestley suggested, it must have been because a servant had played a cruel trick on him and urinated into the water he'd brought to fill Nooth's carbonation apparatus. Priestley had no foundation for this accusation, and he eventually gave up his attack on Nooth and conceded that Nooth's apparatus was superior to his own.

Later, Swiss inventor Johann Jacob Schweppe scaled up the apparatus, making carbonated water available to all. Today, the debate over the health benefits of such waters continues. But the financial benefits are beyond dispute — the lucrative soda pop industry relies on artificially carbonated water.

40. Ogden Nash, the clever poet who treated us to gems like "Candy is dandy but liquor is quicker," once pondered, "Would you be calm and placid / If you were full of formic acid?" To what was he referring?

Ants. These amazing creatures aren't exactly filled with formic acid, but they do produce it for use as a chemical weapon.

If you've ever experienced the burning pain of a fire ant sting, you've experienced formic acid. And you've undoubtedly learned that fire ants are not to be toyed with. Boys of Brazil's Maué tribe could probably tell us a few things about formic acid. As a coming-of-age test, they have to thrust their arms into sleeves stuffed with ferocious ants again and again until they are able to endure the pain without betraying emotion. When a boy has reached this point of endurance, he is considered to be a man, and his elders will permit him to marry.

But you don't have to thrust your arm into a sleeve filled with fire ants to be affected by formic acid. Ants are a natural source of acid rain; they actually account for most of the acid rain in the Amazon. It is estimated that ants release about twenty billion pounds of formic acid worldwide per year.

Formic acid isn't always a problem. People have been putting it to commercial use ever since it was first isolated in the seventeenth century by distilling the residue from ants crushed in a mortar. The dyeing and tanning industries use it, and formic acid is the active ingredient in solutions used to remove scale from the inside of kettles. But animal rights activists needn't worry — the commercial product is not made by distilling ants. It is synthesized from carbon monoxide.

41. Where did the ancient Greeks think the entrance to hell was located?

Does hell exist? The ancient Greeks certainly thought so. They even knew where the entrance to the underworld was to be found: right beside the Temple of Apollo in Pamukkale, in what is now Turkey.

The gateway looked like the entrance to a cavern, but this was no ordinary cavern. No animal or man who wandered into its misty interior ever returned. Now we think we know why. Subterranean hot streams penetrate the ground surrounding the cavern. As the streams flow over deposits of limestone — or calcium carbonate — the water picks up carbon dioxide gas. Sort of a natural carbonation process. Then, as the carbonated water reaches the cavern, the pressure is released, and the gas escapes. It's kind of like opening a bottle of pop. Since carbon dioxide is heavier than air, it pushes the air out of the cave, so any person or creature who enters is quickly overcome by oxygen deprivation.

The Pamukkale cavern wasn't the only place where carbon dioxide wreaked havoc with human lives. On August 21, 1986, a terrible natural-chemical accident occurred in Cameroon, Africa. Within hours, thousands of animals and people had died in the area around Lake Nyos. It was as if some celestial plague had struck people dead in their tracks, but the dead showed no outward signs of disease. The culprit turned out to be carbon dioxide gas — the gas that bubbles from our soft drinks and that we exhale after every intake of breath. Where did it come from? The bottom of the lake.

The devastating accident was probably due to underground volcanic activity, which caused a carbon dioxide buildup. The gas bubbled suddenly to the lake's surface and spread quickly to neighboring areas. Because carbon dioxide is denser than air, it hugged the ground and flowed down into valleys. The cloud traveled more than fifteen miles from the lake, and at some points it moved fast enough to flatten vegetation, including a few trees. About 1,700 people died of suffocation, and nearly a thousand more were hospitalized.

This type of disaster may well happen again. In all likelihood, carbon dioxide is building up under the lake again. Hell on earth, indeed.

42. What chemical is useful to people who own swimming pools and to French vineyard operators?

——

Copper sulfate. Algae often invade swimming pools, turning the water greenish and cloudy. By adding a very small amount of copper sulfate to pool water, we can clarify it and make the pool more inviting. There is another benefit: by using copper sulfate, we reduce the amount of chlorine or bromine needed to disinfect the pool.

Copper sulfate at such dilutions is an extremely safe substance, and the Pest Management Regulatory Agency of Health Canada has given it a clean bill of health. This agency also regulates a completely different use for copper sulfate — controlling downy mildew fungus, which can devastate grapevines. Downy mildew flourishes in wet weather, showing up initially as a downy patch on the bottom of the leaf. If the rain persists, the fungus establishes itself, eventually destroying the crop; it can even carry over into the next season.

In the 1860s, Pierre-Marie-Alexis Millardet, a professor of botany at the University of Bordeaux, discovered that a mixture of copper sulfate and lime effectively eliminated the fungus. How did he discover this? Farmers had long been spraying their vines with the stuff to render the grapes unappetizing to thieves. Millardet noted that the sprayed vines exhibited no fungal growth. The copper-sulfate–lime concoction came to be known as Bordeaux mixture, and it was the first commonly used fungicide.

Because it is made from naturally occurring minerals, organic farmers use the mixture — despite the fact that toy companies have judged copper sulfate too dangerous to include in their chemistry sets. Growing beautiful blue crystals of copper sulfate was once a standard school science activity, but these days students with inquiring minds have to purchase copper sulfate at health food stores, where it can legally be sold as an organic fungicide.

43. We hear a great deal about lycopene, a compound that can reduce a man's risk of getting prostate cancer. What raw food is the best source of this substance?

———

Watermelon. Not what you thought! Most people immediately think of tomatoes when the subject of lycopene comes up.

Indeed, the red pigment occurs in tomatoes. When it's actually inside the tomato, protecting the fruit from disease and sun damage, this molecule is in a form called trans-lycopene. But in order for our bodies to absorb it well, it needs to be in a form called cis-lycopene. To effect this conversion, tomatoes must be cooked.

The only difference between trans and cis is the shape of the molecule. When lycopene changes into the cis form, it can fit into a receptor on a cell. There are many different types of receptors on cells, and each receptor can receive only molecules with very specific shapes. Cis-lycopene has a receptor on cells that it can call home; trans doesn't. So trans passes right through the digestive system, while cis hooks up and heads into the blood stream, where it goes to work.

Watermelon contains more cis-lycopene, so we don't have to cook it. In a U.S. Department of Agriculture study, twenty-three healthy volunteers drank servings of either watermelon juice or tomato juice containing twenty milligrams of lycopene. In each case, blood lycopene concentrations doubled: that is, they were twice as high as concentrations found in people with low-lycopene diets. Obviously, watermelon isn't the nutritional lightweight most people assume it to be. Anyone for a vodka and watermelon juice?

44. What is the connection between "tangerine trees and marmalade skies" and the morning glory?

———

The Beatles sing of "tangerine trees and marmalade skies" in "Lucy in the Sky with Diamonds," one of their greatest hits. The first letters of Lucy, sky, and diamonds were rumored to stand for LSD, the famous hallucinogen with which the Fab Four undoubtedly had some experience.

LSD, or lysergic acid diethylamide, does not occur in nature, but other, closely related compounds do. The seeds of the morning glory plant, for example, harbor the hallucinogen lysergic acid amide, which is less potent than LSD. The ancient Aztecs prepared a beverage, known as *ololiuqui*, from the seeds of this plant and used it in religious ceremonies. A sixteenth-century Spanish missionary became convinced that the Aztecs were in contact with the devil, observing that, after they'd imbibed *ololiuqui*, they would be "deceived by various hallucinations which they attribute to the deity which they say resides in the seeds."

LSD was synthesized in 1938 by the Swiss chemist Albert Hofmann, who had been studying hallucinogenic compounds found in the ergot fungus, which grows on rye. He attempted to synthesize chemical derivatives of these substances in the hope that the results of his experiments would have some pharmacological use. During his period of experimentation, he accidentally ingested a tiny amount of LSD, probably because he neglected to wash his hands. Later, Hofmann described his remarkable experience: "I sank into a not unpleasant intoxicated condition...I perceived an uninterrupted stream of fantastic pictures, extraordinary shapes with intense, kaleidoscopic play of colors." But, trying a higher dose, Hofmann discovered LSD's downside, describing what happened in his 1980 book LSD: *My Problem Child*: "A demon had invaded me, had taken possession of my body, mind, and soul...I was seized by a dreadful fear of going insane."

Today, we know that large doses of LSD (which are still in the milligram range), interfere with the activity of serotonin, an important neurotransmitter, and this can have catastrophic effects. It's best to limit one's experience with LSD to listening to "Lucy in the Sky with Diamonds."

45. A sixty-seven-year-old man suffering from progressive muscular weakness went to a Taipei hospital. Until then, aside from somewhat elevated blood pressure and an enlarged prostate, he'd been in good health. The only abnormality doctors could find was a very low level of potassium in his blood. The man revealed that he had been taking a Chinese herbal remedy for his prostate. What natural product did this remedy contain that could cause the muscle weakness?

Licorice root. Licorice has a long history in Chinese herbal medicine as a general healing agent. While accounts of its miraculous curative powers are highly suspect, glycyrrhizic acid, its active element, certainly does have physiological effects.

Licorice was one of the first substances that Western physicians used to treat Addison's disease, an ailment of the adrenal glands. Addison's sufferers don't produce enough cortisol, a hormone that helps regulate blood pressure and water retention. Glycyrrhizic acid interferes with an enzyme that normally breaks down excess cortisol in the kidneys; when this enzyme is rendered inactive, cortisol levels rise, and this can cause hypertension. Furthermore, high cortisol levels cause retention of sodium and excretion of potassium. Sodium retention, in turn, leads to excess water retention, and potassium deficiency leads to muscle weakness.

So, what did the doctors advise the gentleman from Taipei to do? They told him to stop eating licorice and take potassium supplements. In two weeks, the man lost almost five pounds because he was no longer retaining water; his potassium and blood pressure returned to normal. All signs of muscle weakness disappeared.

Physicians and patients need to be aware that licorice root lurks in a number of herbal therapies. Incidentally, there is no evidence to support the notion that licorice is beneficial in the treatment of benign prostate hypertrophy.

46. What is the connection between Frankenstein and a frog's leg?

Mary Shelley's Frankenstein is a classic of English literature. But, like Stephen Hawking's *A Brief History of Time*, it is one of those books that everybody knows about but few actually read.

Mention Frankenstein, and what springs to most people's minds is Boris Karloff's portrayal of the monster. That's because they've skipped the book and seen the movie, which is a true Hollywood-style horror story. In writing her novel, Mary Shelley did not intend to scare her readers — what she penned was a work of science fiction that explored the consequences of allowing science to run amuck. But where did young Mary, who wrote the classic when she was only eighteen, get the idea of creating life in the laboratory?

Christopher Goulding, a researcher in English literature at the University of Newcastle, in England, thinks he has the

answer. In a paper published in the May 2002 issue of the *Journal of the Royal Society of Medicine,* he argues that James Lind, a nineteenth-century Scottish natural philosopher, was the model for the fictional Victor Frankenstein. What evidence does Goulding present? Mary's husband was the famous English poet Percy Bysshe Shelley. He was educated at Eton, Goulding tells us, and his mentor was Dr. James Lind. Passionate about science in general, Lind was particularly drawn to a new area of research, called galvanism. Luigi Galvani of Italy had made the stunning observation that static electricity could make legs removed from a dead frog quiver as if they were alive. The electricity was generated by a Wimhurst machine, a device comprised of two disks that rotate in opposite directions and rub against metal brushes. Lind used one to re-create Galvani's experiments, becoming the first British scientist to do so. He infected his young protégé with his enthusiasm for science, as evidenced by the fact that Percy's room was filled with scientific apparatus.

It's not surprising, then, that after Percy moved to London and married Mary, he dragged his new wife to a public lecture on galvanism, during which the lecturer employed a static electricity machine to "animate" frog's legs. This demonstration of galvanism had quite an impact on Mary — she even dreamed of a stillborn baby being brought back to life with electricity. The stage was set for the creation of Dr. Frankenstein and his monster. In the book, Mary gives no details about the science the doctor exploits to bring his creature to life, but it's a good bet that Dr. James Lind's experiments were in the back of her mind. Lind may have been the real-life Frankenstein. Now…go and read the book!

47. A popular story describes how, in 1638, the Countess of Chinchon, wife of the viceroy of Peru, was cured of a disease then known as tertian fever. The remedy was the powdered bark of a tree. What do we call this fever today, and what is the healing substance in the bark?

The fever is malaria, and the healing substance is quinine. The tale of the discovery and use of quinine as a treatment for malaria is recounted in many a text. But, while it is a romantic story, it just isn't true.

Here is a typical version: Lady Ana de Osorio, Countess of Chinchon, married to the viceroy of Peru, was stricken with malaria. She was at death's door when the Spanish governor of a town in the Andes appeared at her bedside with some powdered tree bark. The natives had told him that the bark would cure the fever, and it did. Later, the countess took the remedy home to Spain, thereby introducing the first effective treatment for malaria to Europeans at a time when tertian fever was rampant on the Continent.

This is the story that Swedish botanist Linnaeus heard in 1735 from explorer and physicist Charles Marie de la Condamine, who had sent Linnaeus some samples of the bark for classification. Based on this account, which had already appeared in written form as early as 1663, Linnaeus named the tree cinchona to commemorate the countess's contribution to public health. In 1874, Clements Markham, who later became president of the Royal Geographical Society, wrote his own version of the story after researching the details on a visit to Spain. Medical texts have since uncritically repeated this tale. And that is just what it is. A tale.

Historical records indicate that the countess died in 1625, having never even visited Peru. Her husband married again, and he took his second wife to Peru, but she didn't contract malaria. We know this because the viceroy's secretary kept detailed records of everything that happened in the household, including the fact that the viceroy himself endured several bouts of malaria. There is no mention in the records of pulverized bark.

Historian A. H. Haggis finally wrote the true story in the 1940s, but the romanticized version still crops up. So, who introduced quinine to Europe? The Jesuits did. They learned about it from the natives — at least that part of the story is true!

48. In some U.S. states, capital punishment is carried out with poisonous gas. The substance is actually generated in the gas chamber after the executioner slips it, in pellet form, into a clear liquid. What is the gas, and what chemicals are used in the reaction?

––––

The gas is hydrogen cyanide, and it is generated when solid sodium or potassium cyanide is combined with concentrated sulfuric acid. Hydrogen cyanide possesses the somewhat pleasant odor of bitter almonds, but that's where the pleasantness stops. The stuff kills very quickly. A concentration of over 250 parts per million in air is immediately fatal.

Cyanide shuts down the body's ability to produce energy through the normal process of respiration. The body requires energy to perform all of its vital functions. In respiration, oxygen combines with fuel, mostly glucose, to produce water, carbon dioxide, and energy. In order for this to happen, cells have to make use of an enzyme called cytochrome oxidase. It is this enzyme that is inhibited by the presence of hydrogen cyanide. The supply of oxygen to the body's tissues may still be plentiful, but the cells cannot utilize the oxygen effectively. The Nazis used hydrogen cyanide, under the name of Zyklon B, in their gas chambers during World War II.

Exposure to hydrogen cyanide can occur outside of the gas chamber as well. When substances that contain nitrogen — like wool, silk, nylon, or polyacrylonitrile — burn, they release the gas. In intense heat, carpeting, for example, releases hydrogen cyanide. Firemen have to contend with this problem when battling fires in modern buildings full of synthetic materials. In an intense fire, though, hydrogen cyanide itself will burn to produce carbon dioxide, nitrogen dioxide, and water, which are of no concern.

49. Virtually everyone is aware of the connection between dietary cholesterol and heart disease. But many are surprised to learn that cholesterol is a common ingredient in cosmetics. Why is it used for this purpose?

———

Many cosmetics, from lipsticks to moisturizing creams, contain emollients, whose main function is to keep moisture in the skin from evaporating. Water loss reduces the suppleness and smoothness of skin, leading to a dry, scaly texture.

Emollients are chemicals that coat the skin, preventing moisture from passing through. Oils are very effective emollients, but most people are put off by their slippery feel. The challenge for cosmetics manufacturers is to find substances that prevent moisture loss but do not make the wearer feel like a greaseball. Cholesterol meets this challenge. It is readily absorbed into the surface of the skin, and it doesn't feel oily.

But there's a problem. Cholesterol is found only in animals. The most widespread means of procuring cholesterol for commercial purposes is to extract it from the spinal cords of cattle or from lanolin, the natural grease found on wool. These days, people are concerned about products derived from animals — and particularly from cow brains and spinal cords — because they fear that these may transmit bovine spongiform encephalopathy (BSE), better known as mad cow disease.

The prevailing opinion is that BSE was originally caused by feeding animal products, including those derived from sheep, to cattle. When cholesterol is extracted from spinal cords or wool, there is always a chance that it will be contaminated with trace amounts of other animal products, perhaps even the special proteins, known as prions, that scientists have implicated in BSE. The risk of anybody contracting mad cow disease from cosmetics borders on zero, but cosmetics producers have to contend with the consumer perception that animal-derived ingredients are undesirable. For this reason, they are now seeking nonanimal sources of cholesterol.

Chemists at Sigma-Aldrich have developed a process to synthesize cholesterol from plant sources. Some plants, such as the Mexican yam, are excellent sources of specific steroids, which, through a sequence of chemical reactions, can be converted into cholesterol. Cosmetics manufacturers can use this cholesterol in their products, and pharmaceutical researchers can exploit its growth-promoting properties in cell culture. The development of a synthetic process to produce cholesterol from raw materials in plants is another example of problem solving through the appropriate use of chemistry.

50. Kristine Petterson has written a delightful little book of experiments for children entitled *Icky Sticky Foamy Slimy Ooey Gooey Chemistry*. One of the more endearing experiments is called "brain sludge," and it involves mixing milk with a chemical to form "quivering white globs of goo." What is the chemical, what is going on?

––––

The chemical that enables you to "gross yourself out" is simply vinegar. The instructions for the experiment are straightforward: half-fill a test tube with vinegar, fill an eyedropper with milk, insert the eyedropper into the test tube, and slowly squeeze the milk out near the bottom of the tube. Watch the disgusting "brain tissue" form.

The author encourages budding scientists to pour the quivering brain sludge over their hands and enjoy the sensation. But if you have any real brains, you won't do it. Actually, I have to say that, in the proper context, this is a wonderfully instructive little experiment.

"Brain sludge" forms when proteins precipitate from the milk. Proteins are composed of long chains of amino acids, which are curled up into little balls called micelles. These have roughly the same density as the surrounding liquid — water — so they remain suspended. When these proteins come into contact with acids, they uncoil and stretch out. The stretched-out molecules then intertwine with each other to form globs that no longer have the same density as water, so they sink to the bottom of the tube.

While the sludge may look totally unappetizing, if you just add a little salt, you've got cottage cheese. Long before we had supermarkets, people had to make their own cheese, and they did so by acidifying milk. Now, wouldn't "cheese making" be a more interesting term to describe this experimental activity than "brain sludge" making?

51. An Italian lady showed up at a clinic complaining of generalized itching and lip swelling about thirty minutes after making love with her husband. The reaction disappeared after treatment with cetirizine (Reactine). Tests showed that she had no food allergies. The lady mentioned to her doctors that her husband had recently been diagnosed with gingivitis. Can you figure out what happened?

———

About two hours before making love to his wife, the husband had taken an antibiotic, bacampicillin. The trace amounts of this compound transferred to her lips as they were kissing were enough to cause an allergic reaction.

Physicians suspected the reaction after the forty-five-year-old woman revealed that four years earlier she had experienced similar symptoms after taking bacampicillin. On that occasion, her doctor had treated her successfully with cetirizine and an intravenous injection of hydrocortisone, suggesting an allergic reaction. In order to confirm the diagnosis, the clinic doctors subjected the lady to a battery of allergy tests to rule out reactions to food or inhaled substances.

Then they gave the husband either a placebo or bacampicillin over the course of several days. Two hours after ingesting the medication, they instructed, he must kiss his wife. Why two hours? Because that is roughly the length of time it takes for a drug to show up in salivary secretions. Just as they had suspected, about twenty minutes after kissing her husband on the days he'd taken the antibiotic, the lady felt her mouth begin to itch and saw wheals appear on her face and arms. Within an hour of ingesting the antihistamine cetirizine, the symptoms disappeared.

But why was the husband prescribed the antibiotic? Gingivitis is an inflammation of the gums caused by the accumulation of bacteria. It is almost always the result of inadequate brushing and flossing, which lead to a buildup of plaque along the gum line. Plaque is the soft, sticky film composed primarily of bacteria that hardens into tartar. Gingivitis is a danger not only to the teeth but also to the heart. When the gums bleed, bacteria can enter the bloodstream and provoke inflammation in the coronary arteries, and this can lead to heart disease.

52. What is emu oil?

Look up in the sky...it's a plane...it's a bird...Actually, if you want to see this bird, you have to keep your gaze at ground level. It's an emu. A big bird. And it doesn't fly. It walks on the ground.

The emu, native to Australia, can grow to about six feet tall, and it's one strange-looking bird. But it's not as strange as some of the health claims that people have made for its oil. For centuries, Australia's Aborigines have used emu fat for medicinal and cosmetic purposes. In the 1980s, emu fat, in the form of purified oil, caught the imaginations of others. The oil was widely advertised, and believers offered testimonials to its amazing healing properties. Dozens of conditions — ranging from acne and arthritis to eczema and hemorrhoids — responded to emu treatment, or so some people insisted. A pharmacist even maintained that by rubbing emu oil on his bald head he had regenerated his hair! Individuals suffering from shingles and carpal tunnel syndrome stated that their symptoms had improved. And on it goes. Emu oil makes cuts and burns heal faster, and it takes the sting out of fire ant bites. Ninety-five percent of National Basketball Association teams use emu oil to treat injuries.

It all sounds very interesting, but it's somewhat puzzling. Especially when we consider the chemistry of emu oil. Just like any other fat, it is composed of a variety of fatty acids, mostly oleic, with smaller amounts of palmitic, linoleic, and stearic acids. There is nothing special here. Of course, it also contains compounds other than fats. We know that it has a variety of terpenes, sapogenins, and flavones, and, if emu oil really does have activity, then the magic must reside in these compounds. Indeed, researchers at Women's and Children's Hospital in Adelaide, Australia, are looking into this. If they manage to identify the active anti-inflammatory ingredient, then perhaps emu oil can be produced in standardized versions.

What we do know is that the oil penetrates skin smoothly, and it's a good emollient — that is, it prevents moisture loss and makes the skin more supple. But there are many other oils that do this too, and they cost a lot less.

Controlled studies of emu oil have never been conducted. The oil has inspired an array of anecdotal accounts — as have substances ranging from aloe to vitamin E to actual snake oil — and these tend to focus on its ability to heal minor wounds and relieve pain when rubbed on joints. These anecdotes may sound impressive, but most are meaningless. For example, one lady says her headache disappeared less than half an hour after she massaged a couple of drops of emu oil into her temples. It's hard to imagine how any component of the oil could be absorbed into the blood vessels, travel to the brain, and then influence the dilation or constriction of blood vessels. But it's easy to imagine that the lady imagined her headache disappeared because of the oil. Or that it merely resolved itself. There is certainly no harm in rubbing a little emu oil into the skin, but we really need more evidence before we bite on the healing properties of emu oil.

Biting on emu meat is not a bad idea. It's lower in fat than beef, and it has a very interesting taste. And there may even be another use for emu. Ivan Durrant, an Australian artist, says he's noticed a most unusual emu effect. Durrant carves patterns into the shells of emu eggs, causing a lot of eggshell dust to fly around. It seemed to him that when he licked the dust off his fingers, his sex drive escalated. Researching the phenomenon further, he learned that a quarter of a teaspoon of powdered emu eggshell was enough to stimulate him for at least two days. Sounds about as reliable as curing baldness with emu oil.

53. What is common to hydrazines in mushrooms, estragole in basil, allyl isothiocyanate in horseradish, and myristicin in carrots?

They are all potentially toxic. The first three compounds can produce tumors, and myristicin can cause liver damage and hallucinations. But before we swear off them, let's throw a little dose of reality into the pot.

These compounds are indeed toxic — if you're a laboratory rat. But just because we can provoke adverse reactions in lab rats, it doesn't mean that humans consuming tiny doses are at risk. The human liver is a wonderful detoxifying organ, and it has available to it a variety of enzymatic reactions that break down toxins. This is not surprising, because our food is filled with toxins.

Why are these toxins present? Because, over the millennia, plants have evolved elaborate defense systems against insect predators. The natural pesticides they produce, such as the ones in mushrooms, basil, and horseradish, constitute the vast majority of pesticide residues we ingest through food. For every gram of synthetic pesticide that gets into our bodies, we ingest roughly ten kilos (twenty-two pounds) of natural pesticide.

We don't worry about these potential poisons because the dose is too small. As Paracelsus, the great alchemist of the Middle Ages, put it, "Only the dose makes the poison." Most people don't fuss about the fact that their mushrooms harbor a known carcinogen. Why not? Either because they don't know it's there, or because they believe that naturally occurring carcinogens are somehow different from synthetic ones. Yet, if a food additive turns out to have carcinogenic properties, the same people will raise a great hue and cry. Natural toxins and synthetic toxins are not different. Both are dangerous in high doses and safe in small amounts.

In fact, processed foods are better regulated in this area than are natural ones. Safrole, for example, is a compound found in oil of sassafras, isolated from the bark of the sassafras tree. Health food advocates often recommend sassafras tea as a restorative. At one time, manufacturers used safrole as a root-beer flavoring, but they stopped when safrole was found to cause cancer in test animals. Processed foods, you see, are not allowed to contain known carcinogens — but natural foods are.

54. What is common to Miss Piggy and the World Cup?

Polyurethane. The ball used at the World Cup and the glamorous Miss Piggy, everyone's favorite sow, are both made of this polymer. That is, as long as we are talking about the original Miss Piggy, not the clones on sale in toy stores.

Our story begins back in the 1930s, when Wallace Carrothers at DuPont invented nylon by linking small molecules together to make a giant molecule called a polymer. Carrothers patented his process, but that didn't stop other chemists from seeking other types of molecules that could be linked together in a similar fashion.

In Germany, Otto Bayer attempted to circumvent the nylon patent, and in doing so discovered that molecules called diisocyanates reacted with another class of substances called diols to make polymers. Furthermore, Bayer found that, if the reactants were contaminated with a trace of water, the water would react with some of the diisocyanate to form carbon dioxide bubbles. As the bubbles escaped, they caused the plastic to foam.

Polyurethanes can have diverse properties, depending on which diisocyanate or diol is used and on whether water is excluded. Foams can be hard — like insulating material — or soft — like car seats or Miss Piggy. (Truth be told, only the early versions of Miss Piggy were made of polyurethane. As her popularity grew, she had to be mass-produced, so her makers switched to natural latex foam rubber, which they could mold readily.)

In the absence of foaming, we can produce a material hard enough for roller blade wheels or soft and pliable enough to mimic leather. In fact, at the New York World's Fair of 1964, DuPont proudly introduced shoes made of Corfam, a polyurethane-based synthetic leather. The shoes were durable and waterproof, and they could be cleaned with a moist cloth. Best of all, Corfam was porous, so the shoes breathed like leather. Within a year, seventy-five million pairs of Corfam shoes had been sold, and the leather industry stopped breathing. But the Corfam fad didn't last. Despite the material's porosity, many complained of hot, sweaty feet, and furthermore, women didn't like the idea of shoes that lasted forever — they wanted new styles every year. So, Corfam became DuPont's Edsel.

Yet polyurethane leather has made a comeback. At the World Cup, the polyurethane ball has replaced the traditional leather ball. The outer skin is made of multiple layers of polyurethane,

DR. JOE AND WHAT YOU DIDN'T KNOW

and it covers a layer of foamed polyurethane. The ball doesn't absorb water like leather, and its behavior is highly predictable. It is the result of cooperative research between Adidas and the Bayer chemical company. Bayer, of course, is also known for aspirin, which soccer players who get headaches from heading hard balls often take. The new ball has shock-absorbing properties — so no more headaches.

55. A popular commercial product was first made in the nineteenth century from beef tallow, skim milk, cow's udder, pig's stomach, and sodium bicarbonate. What was it?

Margarine. Emperor Napoleon III offered a prize to anyone who could find "a suitable substance to replace butter for the navy and the less prosperous classes." Chemist Hippolyte Mège-Mouriès, already at work on this very project, quickly submitted his entry. He won — although it came as no great surprise to anyone, because his was the only entry.

The inventor had noted that cow's milk contained fat, even if the animal from which it was taken was malnourished and losing weight. Concluding that the fat was body fat, he chopped some suet, minced in some animal stomach, and cooked the mixture in slightly alkaline water to get "butter." After tasting the result, he decided it needed more cow flavor, so he added a little chopped cow's udder.

Margarine was patented in 1871; by 1880, it was being widely produced, because it was much cheaper than butter. The big turning point in margarine manufacture came in 1905, when scientists developed the process of hydrogenation. Manufacturers could now treat vegetable oils with hydrogen gas, thereby transforming the oils into a solid that had the texture of butter. The dairy industry swung into action.

The dairy lobby urged governments to restrict margarine sales, to levy taxes on margarine, and to prevent margarine manufacturers from marketing products dyed to look like butter. Margarine producers responded by selling their product together with a packet of yellow dye, which consumers were instructed to blend in themselves.

Despite these roadblocks set up by the dairy industry, today we consume roughly three times as much margarine as butter. And why is it called margarine? Mège-Mouriès named it after the Greek word *margarites*, meaning "pearl," because his concoction had a pearly appearance during the mixing process.

56. Why does a Swiss army watch glow in the dark only sometimes?

Luminous paints, such as those used on watch dials, contain chemicals that absorb light and reemit it over an extended period. The phenomenon was first noted in the 1500s, when stones discovered near Bologna were found to glow after exposure to sunlight. Some would glow in the dark for years after they'd been heated intensely in the presence of carbon black.

As you can imagine, this quality was very attractive to alchemists and others who dabbled in mysterious arts. The stones turned out to be barite, or barium sulfate. When heated with carbon, the barium sulfate converted to more luminous barium sulfide. The discoverers of these rocks dubbed them lapis solaris, or sun stone. Since then, many other light-storing compounds have been discovered. One of these, calcium sulfide, is commonly used to create theatrical effects and luminous dials.

Whether the watch dial glows or not depends on the intensity of the light to which it has been exposed during the day and on the amount of time it has spent in darkness. The glow fades as time passes. This phenomenon is amenable to many applications. A few years ago, glow-in-the-dark boxer shorts emblazoned with the word *yes* were big sellers. Then there are glow-in-the-dark condoms, which gleam for a full fifteen minutes after a few seconds of exposure to bright light, giving the expression "Rise and shine!" a whole new meaning. And, finally, there is the glow-in-the-dark toilet seat. Its inventor, a fourth-grader, won a prize for it at an inventors' competition. The judges were impressed by the idea that it could keep people from stumbling around their dark bathrooms in the middle of the night. Sounds like a great idea — the glowing seat helps men to find their target and women to see whether the men have returned the toilet seat to the down position!

57. What interest do scientists who study volcanoes and those who study wine making share?

———

They are both interested in the chemistry of sulfur dioxide. Let's deal with the volcanologists first.

Roughly 300,000,000 tons of sulfur dioxide are released into the atmosphere every year, and about half this amount comes from volcanoes. The rest is a result of human activity. Whenever coal or petroleum burns, some sulfur — which is always present — combines with oxygen to form sulfur dioxide. No matter what its source, the sulfur dioxide oxidizes further in the air to become sulfur trioxide, which dissolves in rain to form sulfuric acid. Presto, we have acid rain! Volcanologists study

volcanic eruptions to determine how such activity affects the production of acid rain.

But what about vintners? They have used sulfur dioxide since the time of the ancient Romans to control the growth of rogue yeasts in wine. Traditionally, they would burn sulfur near the grape juice to kill undesirable yeasts before adding the yeasts that produce wine. Today, instead of burning sulfur, many vintners use sodium metabisulfite tablets. These release sulfur dioxide as needed.

Some people are sensitive to sulfur dioxide, even in trace amounts, and they can suffer severe reactions, including anaphylaxis. This is rare. Our bodies are actually used to handling sulfur dioxide, because it forms when we metabolize certain amino acids. It is usually detoxified by conversion to sulfates, which we excrete harmlessly.

Due to its versatility, sulfur dioxide is also a widely used food additive. It controls microbial decay and prevents reaction with oxygen, thus reducing the risk of browning. Sulfur dioxide also preserves the vitamin C content of food. So, plenty of scientists are interested in sulfur dioxide for plenty of reasons.

58. In the early 1940s, a joke went around about a farmer who complained that, if his crop didn't get enough rain, he'd only be able to grow Baby Austins. What prompted this joke?

Henry Ford's idea to use materials derived from soybeans to make his cars. Ford grew up on a farm, and early in his career developed an interest in what he called "chemurgy" — the science of finding new industrial uses for crops.

By the 1930s, Ford's researchers had determined that the soybean had an excellent contribution to make to this science because it produced a versatile oil and contained a protein that could be turned into fibrous materials. The Chinese had already established soybean processing, and Ford sent some of his people to China to study their methods. Here they learned how to isolate the oil and the protein; they also learned that the Chinese worked in the nude. This aspect of the technology did not appeal to Ford, but it may have inspired him to make fabric out of soybeans. On special occasions, Henry would sport a soybean-fiber suit.

Ford researchers found more practical uses for the soybean. They mixed the oil into enamel car paint, and they molded soybean meal to make horn buttons, gearshift knobs, door handles, and even acceleration pedals. Robert Boyer, Ford's main soy researcher, made plastic sheets of soybeans that could replace steel. He installed a soy trunk lid on one of Henry's personal automobiles, and the magnate delighted in gathering an audience of skeptics and smashing the lid with an axe.

Then war broke out, and automobile production came to a virtual standstill while car factories cranked out military equipment. The war also triggered a great deal of research into plastics, and by the time it ended petroleum-based plastics were commonplace. They were easier to produce than soy products, and at that point nobody worried about running out of petroleum or the environmental consequences of the new technologies. But, given the current climate, we may yet see the rebirth of the soybean car and suit. And if you grow tired of wearing or driving your purchase, you can always eat it.

59. What do the SPF numbers on sunscreens mean?

Bottles of sunscreen on the drugstore shelf prominently display their sun protection factor numbers. The higher the number, the greater the protection. But what do these numbers actually mean?

It's very simple. If you use a product with an SPF of two, you can remain in the sun without damaging your skin twice as long as you could without any protection. A person using this product who normally burns in fifteen minutes won't start to fry for half an hour. So, anything above SPF fifteen provides substantial protection. And we need protection, because excessive exposure to ultraviolet light is connected to photoaging and skin cancer.

How good is the protection that the active ingredients in sunscreens and sunblocks offer? We know that they absorb ultraviolet light effectively. But are all sunscreens chemically inert on the skin? Perhaps not. When skin absorbs harmful

UV rays, the energy of the radiation has to go somewhere. In the case of padimate O, for example, a common ingredient, UV absorption may result in the formation of hydroxyl radicals, which can damage DNA. Zinc oxide and microfine titanium dioxide may also present problems. In fact, we know that titanium dioxide absorbs UV and generates free radicals. We already use it for this purpose — as a photocatalyst in waste-water treatment.

Another sunscreen, 2-phenylbenzimidazole-5-sulfonic acid (PBSA), may also damage DNA when it is exposed to ultraviolet light. When a mixture of DNA and PBSA is exposed to UV-B, the molecules become excited and cause a break in the DNA double helix. Perhaps the excited PBSA transfers energy to oxygen, producing a highly reactive form known as singlet oxygen, which can damage DNA. While this does happen in the test tube, there is no evidence that PBSA molecules penetrate skin cells to cause similar damage.

Of course, our best option is still to use sunscreens. The good they do far outweighs any theoretical problems they may present. Sunlight can be very damaging. A Swedish study has revealed that people under thirty who expose their skin to sunlamps experience a threefold increase in melanoma. Still, we must avoid becoming paranoid about sun exposure, because it does trigger the formation of vitamin D in the body, and this vitamin is important for the prevention of osteoporosis. It may also inhibit the growth of prostate and breast cancer cells.

Those who live in northern climates, due to the angle of the sun in the winter, produce no vitamin D, so they should take a multivitamin containing at least four hundred international units of the vitamin. Or they can vacation somewhere warm and sunny — as long as they remember to pack the sunscreen.

60. What is the chemical link between launching the space shuttle and welding rails for trains?

———

Both processes rely on aluminum's tendency to react with oxygen. It is due to this reaction that aluminum does not tarnish. Any sample of aluminum will react with oxygen in the air to form a thin layer of aluminum oxide, which adheres strongly to the surface of the metal and protects it from further reaction.

This reaction is exothermic — that is, it releases heat. At room temperature, however, the reaction is slow, and the heat dissipates into the air. But if we heat the aluminum to a high temperature, the reaction proceeds very quickly and releases a great deal of heat. Even if we remove the heat source, the reaction will proceed because the heat it releases is enough to keep initiating the reaction of the as-yet-unreacted aluminum. The oxygen source doesn't have to be the air; it can be a compound that contains sufficient oxygen, such as iron oxide, or rust.

When workers use the thermite welding process on train rails, they place a mixture of aluminum and iron oxide between the rails or in cracks in the rails. Next, they insert and ignite a piece of magnesium ribbon. This produces enough energy to get the reaction going; the aluminum then reacts with the iron oxide, stripping it of oxygen and converting it to metallic iron. So much heat is generated that the iron melts, filling in the cracks or spaces and forming a solid mass with the rest of the iron as it cools.

In the solid fuel boosters of the space shuttle, aluminum powder mixes with ammonium perchlorate, an excellent oxidizing agent (in other words, it's a great source of oxygen). When the reaction begins, aluminum converts to aluminum oxide, which at high temperatures vaporizes and escapes from the rocket engine, providing the thrust needed for liftoff.

During World War II, scientists used the same chemistry to make incendiary bombs. They constructed the devices of aluminum and iron oxide and ignited them with magnesium. But they put them in magnesium casings so that, when the reaction got going, the casings would ignite and generate a tremendous amount of heat.

A knowledge of this chemistry can come in handy, particularly if you tend to get yourself into tight spots, like 1990s TV hero MacGyver. On one occasion, he had to extricate himself from a particularly nasty situation. He found an aluminum can, which he stuffed into an iron pipe. Then he dropped the pipe into a nylon stocking that had been dipped in oil. Igniting the stocking initiated a reaction that produced enough heat to blow up a car battery. In the ensuing mayhem, the clever MacGyver made his getaway. Pretty ingenious.

61. What is the heaviest naturally occurring element?

Uranium, with an atomic number of 92. The atomic number is the number of protons in the nucleus of an atom, and it defines the element. But when there are more than ninety-two protons, the nucleus is destabilized and decomposes to form other elements. We know this because scientists have been able to make such nuclei, and have observed that the nuclei are stable for only a fraction of a second.

How did they make these nuclei? By employing the process that the alchemists of yore sought so valiantly: transmutation. Changing one element into another. But such procedures aren't really part of the realm of chemistry — instead, they belong to the world of high-energy physics, nuclear fusion, and particle accelerators costing millions of dollars. Scientists recently made element 112, for example, by accelerating zinc ions (atomic number 30) to high energies and slamming them into a lead target (atomic number 82). For less than a millisecond, a nucleus with 112 protons actually existed. It decayed almost immediately into other elements, but scientists were able to identify it by its decay pattern.

What is it called? It remains unnamed for now, but eventually someone will come up with an appropriate name. Other elements have been named after countries, as in germanium for Germany; after people, as in mendelevium for Mendeleev; and after planets, as in uranium for Uranus. But naming element 112 may take some time, given the bickering that American scientists, German scientists, and scientists from the former Soviet Union have engaged in over naming the elements above atomic number 103.

There finally seems to be some agreement on the name for element 106; most are willing to call it seaborgium, after American fusion-research pioneer Glen Seaborg. Nielsbohrium, nahnium, and meitnerium are being bandied about for 107 through 109. So, how do we refer to 112 for now? The temporary name is "unnildodecium." But by the time we've said it, it no longer exists.

62. Why does your tongue stick to metal on a cold winter day but not to wood or plastic?

The metal, plastic, and wood are at the same temperature, but the metal feels colder. That's because metal is a better conductor of heat, and it conducts heat away from your body faster than plastic or wood do. This also explains why you can reach into the oven and touch the cake you're baking but not the metal cake pan.

Indeed, it's a bad idea to lick a cold metal surface in winter. The heat from your saliva will transfer so quickly to the metal that the liquid freezes almost instantly, adhering tongue to metal. Should you witness such a calamity, however, use a little ingenuity, like the father of a five-year-old Alaskan boy.

Father and son attended a performance at a local high school. When they left the building, the temperature was well below the freezing point. The child, for some unknown reason, licked a metal handrail, and his tongue and upper lip froze instantly to the metal. His quick-thinking dad saw right away that there was no way to pull his son free without wounding him, so he urinated on the affected area. It worked like a charm. Despite the fact that his tongue had been loosened, the boy was probably speechless.

63. The age of air travel was ushered in by a sheep, a duck, and a rooster. Why?

On September 18, 1783, living creatures other than birds or insects became airborne for the very first time, carried aloft in a balloon of paper-covered canvas. Suspended beneath the balloon, the brainchild of brothers Joseph-Michel and Jacques-Étienne Montgolfier, was a wicker basket sturdy enough to carry three passengers: a sheep, a duck, and a rooster.

The brothers had become interested in air travel while reading Joseph Priestley's *Experiments and Observations on Different Kinds of Air*, and they started experimenting with balloons. In 1782, they constructed a silk balloon and filled it with hot air by burning chopped hay and wool beneath an opening in the silk. They actually thought that the balloon was filled with smoke, not the hot air generated by the fire, and they began to experiment with various substances in an effort to identify

the one that would produce the thickest smoke. At one point, they even tried old shoes and rotten meat. They had arranged to conduct this demonstration at Versailles for the king and queen, but the royal couple never witnessed the show — they had fled the scene due to the horrific smell.

But by the time they launched their first trio of aeronauts, the Montgolfiers had figured out that it was hot air that made balloons rise. The three adventurers returned safely to Earth, although the rooster was the first aeronaut to be injured in the line of duty: he sustained a broken wing when the sheep stepped on him. On November 21, 1783, humans followed in the footsteps of the animals. Two men made a seven-mile voyage in a *montgolfière* — as the balloons came to be called — reaching an altitude of three thousand feet. The age of air travel had begun.

64. Why does ouzo turn cloudy when you add water to it?

Ouzo is savored throughout the Mediterranean region. In Greece, Cyprus, Lebanon, and Turkey, people sit in cafes with small glasses of ouzo and little jugs of water on the table before them. When they pour the water into the ouzo, the ouzo turns cloudy. And they say that, the cloudier it is, the better the ouzo.

The white stuff is a precipitate that comes out of solution when you add water. Ouzo makers take pure alcohol and flavor it with aniseed and other aromatic herbs. Each producer makes a distinct ouzo, because the herbs differ. The aniseed and the herbs contain numerous compounds, many of which are more soluble in alcohol than in water. Since the extraction of the flavor is done with alcohol that's almost pure, many compounds are forced out of solution when the drinker pours water into his or her ouzo glass. It makes sense that the better ouzos turn milkier, given that they contain more flavor components originally extracted by the alcohol.

65. What is the main use of propolis?

———

Propolis is the generic name for the resinous substance collected by bees from certain trees and plants, and it is truly effective for just one purpose. The bees use it to seal holes in honeycomb, preventing intruders from entering the hive.

Not surprisingly, propolis has a long history as a folk remedy. The conventional wisdom, dating back to about 300 B.C., has it that, since bees seek out propolis, so should people. In the seventeenth century, propolis was an official drug, and London doctors dispensed it to their patients. Despite the fact that no evidence existed for its efficacy, it was considered a general cure-all. In the twentieth century, propolis, like many other natural products, yielded its secrets to chemical analysis.

The stuff that bees collect mainly from poplars and conifers is a mix of dozens of compounds, including fatty acids and flavonoids. Scientists have tested many of these for biological activity, and their tests have shown antifungal and antibacterial effects. But this certainly isn't enough to justify the claims made by propolis proponents that the substance has antibacterial effects superior to those of modern antibiotics. Some have also made the unsubstantiated claim that propolis stimulates the immune system. Others recommend using propolis to treat ulcers and skin problems caused by fungi.

You can purchase propolis in lozenge, cream, or capsule form. Studies have shown that the mild antimicrobial effects of these preparations may be due, in part, to the residue of solvents used to extract the active ingredients. In any case, the fact that in laboratory studies propolis components have shown antibacterial, anti-inflammatory, or antioxidant activity is not significant. Virtually all plants contain compounds that exhibit such effects, but there are no clinical studies that prove propolis is effective in treating any human condition. Bees, however, should not be discouraged from collecting propolis — it has a proven history of efficacy in beehive repair.

66. Caves are fascinating geological formations hewn out of limestone. Dinosaurs helped create them. How?

Caves are created when limestone, which is calcium carbonate, dissolves in acidic water. The acid is produced when the carbon dioxide in water dissolves to form carbonic acid. We know that carbon dioxide is present in the air — after all, we exhale it with every breath. It dissolves in rainwater, making it slightly acidic.

But caves form deep in the ground — how does the carbon dioxide get down there? Essentially, it is formed according to the "dust to dust" concept. How is it that the human body turns into a skeleton in the ground? Because bacteria feast on the organic matter, decomposing it into a mixture of minerals, water, and carbon dioxide. Over time, plants and animals — like the dinosaurs that perished millions of years ago — have accumulated in the earth and released carbon dioxide. This carbon dioxide dissolves in underground water, making it acidic. As the acidic water flows through limestone deposits, it slowly wears away the rock, leaving large caverns.

And where does the limestone come from in the first place? From the shells and skeletons of sea creatures. About

300,000,000 years ago, the Earth was covered with oceans. When the sea creatures died, their skeletons accumulated on the ocean floor and gradually turned into carbonate rock, or limestone.

Cave formation continues. Every once in a while, a sinkhole appears, making us painfully aware of this "sudden" phenomenon. Of course, the cave hasn't really formed suddenly. For millions of years, the underlying limestone slowly wears away until the ground can no longer support any weight. It gives way, and houses, cars, and trucks are swallowed up. Almost like being swallowed up by a tyrannosaurus rex. And it's possible that the t-rex — or at least its decaying body — played a part in the disaster.

67. What is the difference between natural vitamins and synthetic ones?

————

It depends on which vitamins we're talking about. First of all, we have to understand that the properties of any substance are determined by the structure of its component molecules, not by its origins.

Vitamin C that is synthesized in the laboratory is composed of exactly the same set of atoms joined in exactly the same fashion as the vitamin C produced by an orange or a rose hip. As far as biological activity goes, the source of vitamin C is irrelevant. The cheapest version is as effective as the most expensive.

The vitamin E story is different. Natural vitamin E, as it occurs in foods such as whole grains, is actually a mixture of eight different but closely related compounds. These compounds have different biological activity, as measured by their ability to prevent reproductive problems in rats. The most potent compound is alpha-d-tocopherol, and it's found in so-called natural vitamin E supplements. Remember, however, that this is only one of the eight naturally occurring components.

Alpha-d-tocopherol can also be readily synthesized in the laboratory. In the course of the synthesizing process, however, alpha-d-tocopherol is produced together with alpha-l-tocopherol, its mirror-image molecule. The latter is not found in nature, and it has less biological activity than the so-called d-isomer. To get around this confusing business, we rate vitamin E supplements in terms of international units (IU) of activity. In other words, the composition of the synthetic and the natural versions may not be identical, but a two hundred IU tablet will have the same level of activity, regardless of its origin.

But we must keep in mind that neither the natural version nor the synthetic one includes all eight vitamin E components, which occur in whole foods. Researchers are now discovering that some of these other components may be important. Perhaps we will soon have supplements that exactly reproduce natural vitamin E. Then the only question left will be whether we should take any supplements at all.

68. What common feature do snails, spiders, and octopuses have?

They all have blue blood. And we're not talking royalty here — these creatures literally have blue blood.

One function of blood is to carry oxygen around the body. The transport system is actually quite complex; it's not just a matter of oxygen dissolving in liquid blood. In humans, a molecule known as hemoglobin carries the oxygen around, and it has an iron atom embedded in its structure. Oxygen binds to this iron atom and travels to cells, where it is released. If there is insufficient iron in the system, a condition known as iron deficiency anemia — essentially a form of oxygen starvation — occurs. The oxyhemoglobin molecule absorbs all colors of light, but it reflects red, and this explains why our blood is red.

Unlike mammals, snails, spiders, and octopuses do not use hemoglobin to transport oxygen. Instead, they rely on a related compound known as hemocyanin. This molecule doesn't have an atom of iron in its middle; it has an atom of copper, which binds oxygen. Hemocyanin absorbs all colors except blue, which it reflects.

While we don't need copper to transport oxygen around our bodies, we do need some copper in our diets in order to utilize the oxygen. Every cell in the body needs oxygen to survive. When the oxygen is delivered by hemoglobin, it is put to use in a variety of metabolic reactions. These reactions are catalyzed by enzymes that require copper to function. The amount of copper we need to consume is quite small — only a couple of milligrams a day, which we find in foods such as oysters, crab, lobster, red meat, nuts, soybeans, and bran; there's even some in our drinking water.

Just because a little copper is necessary for life doesn't mean that more is better. Indeed, since it can displace iron and zinc from critical enzymes, copper can be toxic. In theory, as little as thirty grams of copper sulfate could be lethal, although one would start vomiting before one could ingest such an amount. Incidentally, copper's name derives from the Roman name Cuprum, for Cyprus. This island was a rich source of the metal for the Roman Empire.

69. Many plants produce salicylic acid, a compound that in the laboratory can be converted into acetylsalicylic acid, or aspirin. Why do plants produce this compound?

————

To protect themselves. When a plant is attacked by a fungus or a virus, it musters its defenses, and its defense can involve a variety of proteins that will attack the attacker. The genes that code for these proteins are turned on by salicylic acid, which is synthesized by the plant in response to stress.

We are interested in salicylic acid because it has anticoagulant properties, and some scientists suggest that one reason an organic vegetarian diet is beneficial is that it raises blood levels of salicylates. Scottish researchers have discovered that organic vegetable soups contain almost six times as much salicylic acid as regular vegetable soups. Still, there isn't very much there: 177 nanograms per gram, versus 20 nanograms per gram. Organic vegetables are grown without pesticides, so they have to protect themselves, and they apparently do so by producing more salicylic acid. Perhaps, through genetic modification, we can increase the levels of salicylic acid in fruits and vegetables.

70. The acidity of what bodily fluid is altered by food?

————

The only bodily fluid with a pH — or acidity level — that changes significantly in response to food intake is urine.

Of course, you may not get this impression if you watch late-night TV infomercials or read "alternative medicine" magazines. These paint a horrific picture of the diseases caused by "acidic blood," and they promote devices that manufacture alkaline water to neutralize excess acid. It's all nonsense. Drinking alkaline water does not change your blood's pH level.

The blood is a buffer system, meaning that it maintains its pH level at about 7.4 with very little variation. Indeed, a drop or elevation of 0.4 units creates a potentially lethal situation. Excess base in the blood is quickly neutralized by carbonic acid, which we generate when we inhale carbon dioxide, and the calcium released by our bones neutralizes excess acid.

It's true that certain foods can have an acid or alkaline residue; we can determine this by burning the food to ash. This is a function of the mineral content of the food. The presence of calcium, for example, leads to an alkaline residue, whereas sulfur, found extensively in proteins, renders the food acidic. That's because sulfur can be oxidized in the body, eventually forming sulfuric acid. Such acidity has no effect on the blood, which is buffered, but it can alter the pH of the urine.

Our bodies eliminate minerals through urine, which is not a buffered system. Chlorine, phosphorous, and sulfur in some foods leave an acidic residue; these foods include breads, cereals, eggs, fish, meat, and poultry. Contrary to what many people think, citric acid does not leave such a residue, because our bodies metabolize it completely. Cranberries, however, contain an acid that we cannot metabolize, and it passes into the urine. Under what circumstances is such knowledge important? When someone suffers from kidney stones, which are composed of calcium and magnesium phosphates, carbonates, or oxalates. These salts are insoluble in alkali and are therefore more likely to form when the urine is alkaline.

Kidney stone sufferers may benefit from a diet that produces an acid urine. Uric acid and cystine stones are formed under acidic conditions, and an alkaline-residue diet counters this. Most fruits and vegetables produce an alkaline ash, but corn and lentils are acid-forming. Drinking alkaline water will solve none of these problems. Machines designed to generate water that treats disease by changing the acidity of the blood are a total scam.

71. Why do roosters crow in the morning?

Roosters crow to attract chickens. Chickens have been around for roughly five thousand years, and the general opinion among evolutionists is that, way back when, roosters crowed all the time because, like men, they were ready to indulge in carnal pleasures at any hour. But the crowing sound attracted not only hens but also predators, so roosters took to crowing at low-light times of day — dawn or dusk — when they were harder for predators to see.

Truth be told, some roosters don't restrict themselves to crowing at dawn or dusk, because these days their predators are scarce. We tend to notice the early-morning crowing, however, because there are few competing sounds at that hour.

Incidentally, chickens don't need roosters in order to produce eggs. They lay them all the time — but, of course, these eggs are not fertile and will never hatch. To produce fertilized eggs, a chicken must mate with a rooster. And that's where the crowing comes in.

72. Paul Revere warned American colonists that the British were coming. Today, Revere's gravestone represents a different kind of warning, because pollution has damaged it irreparably. What is the gravestone made of, and what kind of pollution is destroying it?

Revere's stone, located in a Boston cemetery, is made of marble, and it is being destroyed by acid rain. Marble is calcium carbonate, and it reacts with sulfuric acid, the most common acid found in acid rain, to produce calcium sulfate, which then washes away.

This phenomenon is also affecting the Parthenon, the Taj Mahal, the Canadian Parliament buildings, and the Lincoln Memorial. The chambers beneath the Lincoln Memorial have huge stalactites, which formed as the marble eroded and converted to calcium sulfate.

Where does acid rain come from? Most of it is created when we burn fossil fuels, such as coal, to generate electricity. Coal is the product of decaying vegetation, and, since plants contain a variety of sulfur compounds, coal can have a significant sulfur content. When coal burns, sulfur converts to sulfur dioxide, which further reacts with oxygen in the atmosphere to form sulfur trioxide. This, in turn, reacts with water vapor to form sulfuric acid. Ozone, always present in air to some degree, helps convert sulfur dioxide to sulfur trioxide.

Iron and steel producers burn a great deal of coal at high temperatures, generating sulfur dioxide. Nickel and copper producers do the same. Nickel's source is the ore nickel sulfide, which producers must strip of its sulfur content in order to liberate the metal. The sulfur ends up as sulfur dioxide. The world's largest nickel smelter is in Sudbury, Ontario, and, as a result of the emissions, the landscape around it looks like the moon. Indeed, astronauts have used it as a lunar landing practice site. The smelter has a 1,250-foot smokestack, about the height of the Empire State Building, and it emits huge amounts of sulfur dioxide — in compliance with government regulations.

But sulfur is not the only contributor to acid rain. Oxides of nitrogen also react with water to form nitric acid. And where do these come from? Whenever high temperatures and sparks are produced, nitrogen reacts with oxygen (the two gases that are the basic components of air) to form nitric oxide, which then reacts with more oxygen to form nitrogen dioxide. This happens in the internal combustion engines of vehicles, but lightning is also a culprit.

73. What is the difference between green, black, and oolong teas?

It's all a matter of how you process the leaves of the *Camellia sinensis* plant. All teas come from these leaves, but they can be processed in different ways.

To make green tea, the manufacturer exposes the leaves to steam. The heat treatment destroys the enzymes responsible for the browning reaction that occurs in many plant products when their tissues are crushed. This, for example, is what happens when you cut an apple. Compounds called polyphenols react with the enzyme polyphenol oxidase to form large, complex, brown-colored molecules. Apparently, these products of oxidation have antifungal properties. Tea contains a particular variety of polyphenols, called flavonoids, which have received a lot of attention due to their ability to act as antioxidants — that is, as molecules that neutralize tissue-damaging free radicals. So, in green tea, the original flavonoids remain unaltered.

Manufacturers of black tea do not steam the leaves; they crush and ferment them. This releases the browning enzymes, allowing them to go to work. Antioxidant effects are not destroyed, because the products of the fermentation also have free-radical-neutralizing effects.

Oolong tea is only partially fermented, so we can describe it as being halfway between green tea and black tea. Tea's antioxidant effects have been well demonstrated in laboratory studies, but the health benefits of tea drinking are still up in the air. Heart disease and cancer are common in China and Japan, where the average person consumes six or more cups of tea a day. But the lifestyles and dietary patterns of Orientals and Occidentals differ so much that it's hard to pinpoint tea's role in this. Certainly, though, tea drinking is not harmful, and tea's a great alternative to soft drinks.

74. Brown eggs are traditionally favored by Bostonians, while New Yorkers buy white eggs almost exclusively. Why?

————

Brown eggs and white eggs are laid by different varieties of chickens, but there is no nutritional difference between the two. There is a chemical difference in their shells, however.

Brown eggs contain more of a variety of compounds called porphyrins. These actually fluoresce when exposed to ultraviolet light — an interesting phenomenon. The first chickens that settlers brought to New England were of the brown-egg-laying sort; a white egg was likely to have come from elsewhere and was therefore not as fresh. Chickens in the New York area always laid white eggs, so New Yorkers desiring the freshest eggs bought white. Today, of course, we can transport fresh eggs of any color long distances in refrigerated containers, but the tradition of regional preferences has endured.

The U.S. Food and Drug Administration offers this advice for egg consumers. Store eggs at a temperature of forty-five degrees or under. Avoid purchasing eggs that are not refrigerated. Keep eggs in their carton inside the fridge, not on a fridge-door shelf. Consume eggs within thirty days, and always wash your hands after handling raw eggs. When you order a Caesar salad in a restaurant, be sure that it's made with pasteurized eggs. Don't eat raw dough that contains eggs. Scramble eggs past the runny stage; cook over-easy eggs for three minutes per side; soft-boil eggs until the yolk has started to thicken. Do not keep hard-boiled eggs at room temperature for more than two hours.

75. If you spend a couple of hundred dollars for a pound of coffee, you expect something special. What's so special about Kopi Luwak coffee?

Kopi Luwak beans are processed by a special machine. A living machine. The luwak is a species of civet cat found only on the Indonesian island of Java. Like all civet cats, it possesses anal scent glands that secrete an odiferous fluid. In concentrated form, the fluid smells terrible, but when it's diluted it has a pleasant, musky scent, which is sometimes used in perfume manufacture.

Luwaks, apparently, love coffee, but their tastes are very particular. They will eat only the choicest beans. The luwak's digestive system, however, cannot handle the beans very well — a few hours after it eats coffee beans, the luwak eliminates them in a partially digested form. Somehow, the chemistry of the beans changes when they come into contact with the animal's digestive juices. When the partially digested beans are roasted, the coffee they produce is extremely tasty and full-bodied.

One hopes that the enhanced flavor is due to the partial digestion and not to contamination from the luwak's anal secretions. Plantation workers hunt for the "processed" beans, which, after roasting, are brewed into coffee and served in Indonesia's best hotels. These establishments probably don't explain to their customers the origins of the great taste.

76. In 1826, Antoine-Jérôme Balard poured some chlorinated water into some brackish water he'd taken from a salt marsh in southern France. The water turned brown. What had Balard discovered?

Bromine. Of all the elements in the periodic table, only two are liquid at room temperature. One of them is mercury, and the other is bromine. Bromine is a dark brown liquid that volatilizes readily. Chemists use it as a reagent to analyze for compounds that may contain carbon-carbon double bonds. Bromine reacts with these rapidly, and, if its characteristic color disappears, it means that a double bond is present.

Students routinely carry out this reaction in the lab, but few know the interesting history of the element or realize that care must be taken when using it.

Let's go back for a moment to 1826. The place is Montpellier, France, home of pharmacy student Antoine-Jérôme Balard. Professor Joseph Anglada asks him to look into the possibility of isolating useful substances from the residue that's left when seawater evaporates. Balard treats the residue with chlorine, and this yields a dark brown liquid with an intense smell. Its chemical properties are similar to those of chlorine and iodine, but Balard recognizes it as a new element. He calls it bromine, after the Greek word *bromos*, which means "stink."

Today, we produce bromine the same way Balard did. Our major source of the element is in Arkansas, where there are deep brine wells. Chlorine gas is added to the heated brine, and jets of air or steam drive the bromine out of solution. Bromine and its compounds have many applications. We use the element itself to disinfect water, just like chlorine. Silver bromide is sensitive to light, and it's fundamental to photography. At one time, potassium bromide was a common sedative. We still say "Take a bromide" when we mean "Calm down."

I could have used a bromide the day one of the most memorable laboratory accidents of my teaching career occurred. A student was trying to add some bromine to a solution using a pipette. This is a difficult task, because bromine is heavy — about three times as dense as water. In fact, the stuff is so heavy that it will leak out of the pipette. The unfortunate student did not realize this, and a good dose of bromine spilled onto his arm. Bromine is highly corrosive, and bromine burns can be very nasty. Its vapors can even burn the throat. Luckily, the student suffered no permanent damage.

Balard went on to become a professor of chemistry at the École Normale in Paris, a very prestigious institution. There he discovered chlorine monoxide, an excellent disinfectant. But perhaps his greatest contribution to science was his recognition of the potential of one of his students. The student had completed his studies at the École Normale, and he was set to take up a teaching post at Tournon. Balard, however, prevailed upon the administration of the École Normale to offer him a position there. To make such a request on behalf of young graduates was quite unusual. But Balard insisted that the fellow had remarkable research abilities that could not possibly be fostered at Tournon.

Balard turned out to be right. His student became one of the greatest scientists of the era — thanks to stinky old bromine. The student's name? Louis Pasteur.

77. Bags of tiny, peeled carrots are flying off supermarket shelves these days. How are they produced?

Believe it or not, those cute little carrots are not the result of clever crossbreeding or genetic modification techniques. They are the product of some ingenious machinery.

These babies are actually manufactured from regular-sized carrots with a high-tech carrot-sculpting machine. In the factory, workers dump the vegetables into the machine, which automatically peels them and cuts them into rough thirds. Next, it grinds each piece into the shape of a little carrot. That's why you see a green tinge where the carrot top was once attached on only about a third of the carrots.

Actually, there is some carrot crossbreeding going on as well, because manufacturers use only specially developed sweet carrots for this purpose. The little carrots have been phenomenally successful in the marketplace, and that's great, because they're loaded with beta-carotene, the body's precursor for vitamin A, as well as a variety of other carotenoids. Researchers have linked these compounds with lower incidences of heart disease, cancer, and eye problems.

Plant geneticists are working on producing carrots with even higher levels of these nutrients. Beta III, a carrot containing five times as much beta-carotene as a regular carrot, is the first successful venture of this kind. It may turn out to be an important weapon in the battle to reduce one of the developing world's worst nutritional deficiencies — namely, vitamin A deficiency. Incidentally, cooked carrots are even more nutritious than raw ones, because the cooking process breaks down the cell walls, making the nutrients more readily available.

Carrots are even a good source of calcium, so snack away! And don't worry about the white deposit that sometimes appears on the surface of the little bagged carrots. It's just naturally occurring sugar that crystallizes out as moisture evaporates.

78. Jews celebrate Hanukkah by lighting candles for eight consecutive days. What source of light would hold more historical significance for this festival?

An oil lamp that burns olive oil. Here's why. Around 200 B.C., Israel came under the rule of the Syrian king Antiochus. Because Antiochus was bent on obliterating the Jewish religion, Yehuda Maccabee led a revolution against the Syrians, and, with an army of six thousand, he defeated an army of forty-seven thousand; eventually, the Maccabees were able to liberate Jerusalem and reclaim the holy temple.

In order to rededicate the temple, they had to light the menorah. Its flame symbolized the union of the different types of Jews and the eternal presence of God. They had to use the purest olive oil for this purpose, but only a little bit had been

left in the temple, and they would need at least eight days to prepare more. Nevertheless, they lit the menorah, and, miraculously, that little bit of oil burned for eight full days, until the new batch was ready.

Hanukkah, celebrated by Jews around the world, commemorates the victory over persecution. While olive oil would be the most appropriate fuel for the commemorative flame, celebrants light candles instead, for simplicity's sake. The lighting of these candles does more than pay tribute to the Maccabees — it also symbolizes the kindling of the light of knowledge, generosity, and hope and the driving away of the darkness of ignorance.

During Hanukkah, it's traditional to serve foods cooked in oil, like potato latkes. Olive oil is one of the healthiest oils. This monounsaturated fat does not increase blood cholesterol, and it may partially account for the fact that olive-oil-loving Greeks and Italians have a low instance of heart disease (relative to North Americans), even though their diets contain plenty of fat.

79. The year 1828 was a great one for chemistry. In Germany, Friedrich Wohler demolished the division between natural and synthetic chemicals when he converted ammonium isocyanate into urea. In Holland, Conrad Van Houten accomplished something that would have consequences that were almost as far-reaching. What did he do?

Van Houten invented a press to separate cacao butter from cacao beans, thus making the production of chocolate, as we now know it, possible.

The source of all chocolate is the cacao bean. It grows in pods, which are the fruit of the cacao tree. To process them, one must remove the beans from the pods and deposit them in a heap. The beans will begin to ferment. Sugars will convert to alcohol and acetic acid. As the acid concentration increases, the sprouts inside the beans die and release enzymes that break down the proteins and sugars into a host of flavorful compounds with tongue-twisting names like "phenyl acetic acid," "furfuryl alcohol," "dimethyl sulfide," "2-methoxy-4-methylphenol," and "1-methylnaphthalene." Roasting follows, a procedure that causes the cacao beans to turn brown; this is a consequence of the Maillard reaction, by which amino acids react with sugars to form brown-pigmented melanoidins. After roasting, the sprouts, or nibs, are easy to remove. They are ground to produce a substance that hardens to form chocolate liquor.

At one time, the process ended here. One simply mixed the liquor with water to produce a bitter beverage that was marred by a layer of oil that floated on top. This is where Van Houten stepped in. He invented a hydraulic press that squeezed the cacao fat out of the chocolate liquor. The residue of this process Van Houten pulverized to a fine powder. The beverage it produced was still bitter, but Van Houten found that treating the powder with sodium bicarbonate or ammonium hydroxide destroyed the bitter-tasting compounds. A far better drink could be made from Van Houten's powder than from chocolate liquor.

And what happened to the fat that Van Houten's press squeezed out of the chocolate? Some twenty years after Van Houten's discovery, Britain's J. S. Fry created a chocolate bar by adding cacao butter and sugar to chocolate liquor. This was the world's first eating chocolate. Then, in 1876, Henri Nestlé and Daniel Peter found that adding condensed milk to the bar made for a milder flavor, and Milton Hershey devised methods of mass production. Chocolates have been delighting young and old alike ever since.

80. To what chemical does the following poem refer?

___

A mosquito was heard to complain,
"A chemist has poisoned my brain."
The cause of his sorrow
Was paradichloro-
diphenyltrichloroethane.

Paradichlorodiphenyltrichloroethane is DDT. These days, DDT has a bad reputation. Many people associate it with environmental toxicity and human health concerns.

Rachel Carson, one of the world's first influential environmentalists, called it "the elixir of death," yet Winston Churchill spoke of the chemical in glowing terms. Churchill was impressed by the effectiveness of DDT in helping the Allies to control lice that caused typhus, fleas that caused the plague, and mosquitoes that caused yellow fever and malaria. Indeed, when the

Allies entered Naples during World War II, they were confronted with an epidemic of typhoid fever, and they cleared it up with DDT. Epidemiologists claim that DDT has saved about fifty million lives since it appeared, in 1940.

Paul Hermann Müller was carrying out research on insecticides at Geigy Pharmaceuticals in Switzerland when he came across a compound that had first been made in 1874 by Othmar Zeidler, a young chemistry student. It had remarkable insecticidal properties at very low doses, and it was quickly commercialized. Factories churned out about three million tons of DDT over the next thirty years. Müller was awarded the 1948 Nobel Prize for Physiology and Medicine for his contribution to human health.

In some cases, DDT had truly remarkable effects. Ceylon (now Sri Lanka) struggled with a huge malaria problem until its government began spraying homes with DDT in the 1950s. Within ten years, the annual number of malaria cases dropped from 2.5 million to thirty. Then people began to raise concerns. Some insects had started to show resistance to DDT through natural selection. A few members of these species possessed enzymes that could break DDT down into paradichlorodiphenyl-dichloroethene (DDE), which does not kill insects. The insects that survived DDT exposure passed their protective genes on to their offspring.

DDT and DDE also cropped up in the flesh of birds and fish in areas where there had been extensive mosquito spraying. Long Island, New York, was one such area; the first effects were noted there. Some DDT washed into the ocean, where it became concentrated in plankton. Fish that dined on the plankton and birds that dined on the fish began to accumulate DDT and DDE in their fat. Birds started to lay eggs with very thin shells, and their offspring had poor survival rates. Somehow, DDT had affected the birds' estrogen levels. All of this led authorities to phase out DDT. In Ceylon, the malaria rate shot back up to pre-DDT levels.

DDT was banned in North America in 1972, but people continued to worry about its effects after its exit from the marketplace. Due to its persistence in the soil, and consequently in our food supply, human fat still contains DDT. The amounts are very small, but, if DDT does indeed have hormonal effects, even small amounts could pose a risk. On Long Island, the incidence of breast cancer is 6 percent higher than the national average. Could this be because DDT was once sprayed liberally in this community? People actually frolicked in the spray. Some misguided individuals even sprayed their children because they believed that flies spread polio.

Since DDT is fat soluble, it can persist in the body for a long time. The breast is 80 percent fatty tissue, and, since DDT appears to have estrogenic activity, a connection to breast cancer is conceivable. Researchers, however, have not been able to link DDT with breast cancer conclusively. In a typical study, scientists at the Kaiser Foundation Research Institute in California studied three hundred women who had undergone comprehensive physical exams in the 1960s when DDT was commonly used. Eventually, 150 developed breast cancer. Researchers had kept frozen samples of the women's blood, and, when they examined those samples and compared them with samples taken from a group of cancer-free women, they found that the pesticide residues were the same across the board. Still, concerns linger, especially since studies have linked DDT with the emasculation of alligators in Florida.

So, where does all of this leave us? DDT is not a chemical from hell, as some imply, but neither is it totally innocuous. We no longer use it in North America, but no one has offered convincing evidence that our DDT replacement substances are any safer. Insect control is a necessary evil for which we must pay a price. There's no free lunch. But without insecticides, there might not be any lunch at all for much of the world's population.

81. The four bases that we find in DNA and that are responsible for its helical structure are adenine, cytosine, guanine, and thymine. What is the derivation of these names?

———

These days, just about everyone knows something about DNA. The press repeatedly uses the term "template of life," and this is quite accurate because DNA is like a library of information that our cells draw upon to determine which proteins to make.

Proteins make up important structures of the body as well as hormones and the enzymes that govern the body's numerous chemical reactions. DNA is a nucleic acid, composed of pieces called nucleotides, which are strung together like beads on a necklace. Each of these nucleotides harbors a molecule, commonly referred to as a base. The nucleic acid chain is therefore characterized by a particular arrangement of bases, and the sequence of the bases holds genetic information. There are four bases — adenine, cytosine, guanine, and thymine — and they are named for the source from which they were isolated.

And what's the derivation of those names? Adenine comes from the Greek word for gland, because it was first isolated from the pancreas. Thymine was found in the thymus gland, cytosine was isolated from cells ("cyto" refers to cells), and guanine from bird guano — or bird poop.

82. Eva Perón, wife of Argentinean dictator Juan Perón, underwent a hysterectomy in 1951. What was unusual about this operation?

——

Eva Perón was revered by the people of Argentina. Tragically, she developed uterine cancer, and her husband endeavored to keep the diagnosis a secret from her as well as from the people. Unaware of the gravity of the situation, Eva prepared for a hysterectomy. The operation was performed in a clandestine fashion by Dr. George T. Pack, an American, whom Juan Perón had flown to Argentina. Pack entered the operating room after Eva was sedated, and he left before she awoke. If a famous cancer specialist had openly been involved with the case, the secret would have been out.

Eva died the following year of cervical cancer. Her story was brilliantly told by Andrew Lloyd Webber in his classic musical *Evita*.

83. We generally mine salt, but we can also produce it by evaporating seawater from special evaporation ponds. These ponds are often dyed a stunning blue color — why?

——

To increase the rate of evaporation. Just think about how you feel on a hot summer day if you happen to be wearing black. The black absorbs the heat very efficiently, making you feel quite uncomfortable. White reflects heat — this explains why people tend to wear white clothing in the tropics. Blue is also very good at absorbing heat, although it's not as good as black. But there exists no appropriate, economic black dye that we can later remove from the salt residue. We can, however, easily remove the blue dye triphenylmethane from the salt once we've purified it.

And why do we need so much salt that we have to evaporate seawater? This salt is not destined to be sprinkled on our french fries; it is used to produce chlorine and sodium hydroxide, two of the most important industrial chemicals. Passing electricity through salt water converts sodium chloride to chlorine and sodium hydroxide. This process is at the heart of a huge industry — indeed, it consumes more electricity than any other industrial process, save for extracting aluminum from its ores.

Ornamental ponds are sometimes dyed blue as well. This has nothing to do with salt production — it's a measure to prevent algae growth. Algae require sunlight, and the blue dye absorbs the light, keeping it from stimulating algae growth in the pond.

84. What insect has killed more people throughout history than all the wars that have ever been fought?

The flea. In 1330, the bubonic plague — or Black Death — started to spread via trade routes from Asia to North Africa, Europe, and the Middle East. It was caused by a bacterium now known as *Yersinia pestis*. This microbe normally infects rats, and is spread by rat fleas, which also feed on humans.

It's possible that crop failures in China triggered the epidemic. Rats left the fields to seek food in human dwellings, bringing their fleas with them. The fleas bit the human inhabitants, who fell ill and spread the sickness by coughing and expelling infected sputum.

When the bubonic plague hit, its victim would develop a high fever. The heartbeat accelerated, and headache and muscle pains set in. Then the lymph nodes became swollen. Next came

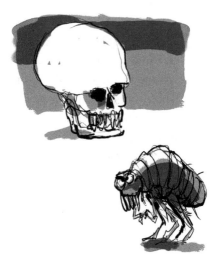

subcutaneous hemorrhaging, which created the black spots for which the horrendous Black Death was named. The victim often developed pneumonia. In a period of twenty years, from 1330 to 1350, the bubonic plague killed about a third of Europe's population.

85. What product contains the following ingredients? Sugar, glucose syrup, water, cornstarch, gelatin, fumaric acid, citric acid, artificial flavor and color, carnauba wax, and maltodextrin.

——

Jelly beans. These little delicacies are made of sugar and cornstarch glued together with gelatin.

We wouldn't have jelly beans if it weren't for all those old horses ready for the glue factory. Gelatin, a mixture of water-soluble proteins, does not occur in nature. It has to be made by boiling animal skin, tendons, ligaments, and bones in water — this may present a problem to people who, for religious or moral reasons, shun animal products. Jelly bean manufacturers add fumaric and citric acids to lend tartness to their products, and they mix in carnauba wax (a tree exudate) and maltodextrin (a starch derivative) to make a shiny coating.

Obviously, jelly beans are not a health food, but they aren't as bad for your teeth as you might think. In fact, potato chips are worse. The sugar in jelly beans is soluble, so your saliva washes it away, but the complex carbohydrates in chips are insoluble — they stick to your teeth, providing food for acid-producing bacteria. Researchers in New Zealand found that the effects on the teeth of jelly beans are no worse than the effects of cereal bars or health bars made with dried fruit.

So, there is nothing wrong with consuming a jelly bean or two on occasion. Like once a week. Certain logging communities in British Columbia celebrate "green jelly bean day" every Saturday. Loggers who are off in the forests all week fly home on weekends with bags of green jelly beans, which they scatter over their green lawns. As the kids hunt for every last sweet treat, Dad goes inside to say hello to Mom.

86. A while back, physicians routinely gave kidney stone sufferers a piece of advice intended to reduce their risk of further attacks. No more. What was this advice, and what changed?

————

Most kidney stones are composed of calcium oxalate, an insoluble substance that forms when soluble calcium compounds combine with oxalate in the kidneys. We get calcium from our diets. We get most of our oxalate the same way, although some can form when we consume excessive amounts of vitamin C.

Logic dictates that reducing our calcium intake reduces our risk of kidney stones, so for many years doctors put their kidney stone patients on low-calcium diets. Then some scientists suggested that this could be the wrong approach. They theorized that, if calcium met up with oxalate in the digestive tract, then it would form calcium oxalate there and prevent the absorption of oxalate into the bloodstream. A landmark paper published in the *New England Journal of Medicine* confirmed this theory. It reported on the discovery that people with a higher calcium intake had a lower risk of kidney stones.

But one conventional bit of advice still holds water. To avoid kidney stones, you should drink plenty of fluids. A Harvard School of Public Health study showed that those who drank water had a reduced risk of kidney stones — the risk declined by 4 percent with each eight-ounce glass of water consumed per day. Coffee lowered the risk by 10 percent, tea by 14 percent, beer by 21 percent, and wine by 39 percent. But apple and grapefruit juice increased the risk; for each eight-ounce glass, the risk rose 36 percent.

87. These days, we are very concerned about lead in our environment. What common product has lead acetate as its active ingredient?

Men's hair dyes of the "cover the gray" variety. Just like the one that hockey star Rocket Richard once advertised on TV.

When the legend was in his seventies, he refereed old-timers' hockey games. He looked pretty good. His hair looked especially good — nary a hint of gray. Of course, it wasn't exactly natural. The makers of a men's hair dye product had hired the

Rocket as their spokesperson. In the television ad, he even got a penalty for "looking so good."

There is some pretty interesting chemistry going on here. Hair dyes guaranteed to banish the gray hairs from your head so gradually that no one will be the wiser actually cover the gray with lead sulfide. Its brown-black color is what does the trick. The active reagents in these hair products are lead acetate and elemental sulfur. Lead acetate is a water-soluble compound, but lead sulfide is practically insoluble. When exposed to the air and to hair, lead acetate reacts with sulfur to form lead sulfide, which precipitates on the hair. Repeated use builds up the lead sulfide, gradually returning hair to a youthful color. At least that's what the ads say. You can usually identify people who have been using the stuff because their hair will have a dark, dull tinge. Still, many think this is better than going gray.

There is, however, one lingering concern about such products. Lead is a highly toxic metal. It poisons the enzymes that make hemoglobin. As a result, a hemoglobin precursor called aminolevulinic acid accumulates in the body and causes toxic symptoms ranging from stomach problems to brain abnormalities. The amount of lead in these dyes is very small — less than 1 percent — and studies indicate that our blood absorbs virtually none of it. But does it contaminate the hands of those who apply it? And what about the excess lead acetate that winds up in our water supply?

Yet another problem arises when people who have already colored their hair with metallic dye use a conventional dye. Manufacturers add hydrogen peroxide to conventional hair dyes to develop the color and to lighten the hair. Many metals catalyze the decomposition of hydrogen peroxide into water and oxygen, and this reaction produces a lot of heat. It can actually cause scalp burns.

88. Diabetics tend to have poor circulation, and they are often afflicted with open sores, called diabetic ulcers. What is the relationship between these ulcers and flies?

Flies lay eggs, which develop into maggots, which hatch as flies. Diabetic ulcers are notoriously hard to heal, but maggots can be very helpful in the healing process.

This therapy has a long history. We know that the ancient Mayans and the Aborigines of Australia applied fly larvae — or maggots — to infected wounds to cleanse them and speed healing. One of Napoleon's military surgeons, Dr. D. J. Larrey, introduced the technique to Europeans. While he was serving in Egypt, he observed that the maggots infesting soldiers' wounds ate only rotting flesh, leaving the healthy tissue untouched. Based on Larrey's observation, many physicians started using maggot therapy, but the practice declined with the advent of effective antibiotics.

Antibiotics, though, don't always work, and the medical community is again interested in maggots. In 1989, doctors at Washington's Children's Hospital were desperate. A seventeen-year-old female patient had an infection on her legs that would not respond to antibiotics. They dressed her legs with fly larvae, and her infection abated. The maggots devoured the rotting tissue, killed bacteria, and promoted healing. For some strange reason, maggots also manage to evade an immune response and do not trigger inflammation. Researchers are now isolating and studying the maggot's secretions, which are responsible for these effects, and their efforts will perhaps give rise to useful drugs.

For now, however, if other efforts fail, doctors can resort to packing their patients' infected wounds with sterilized maggots, which are carefully bred in the lab. And what happens to the maggots once their work is done? The doctor removes the patient's bandage, and the maggots rush out, leaving a clean, healing wound behind them.

89. What is methylcyclopentadienyl manganese tricarbonyl (MMT), and why are people worried about it?

MMT is a gasoline additive that was introduced in the late 1970s, when the additive tetraethyl lead was phased out. Like lead, it reduces the tendency of an engine to knock.

Gasoline with added MMT has a higher octane rating, meaning that it burns more steadily in a car's engine. Controversy has dogged MMT since it first appeared. Gasoline producers, most notably the Ethyl Corporation, which once manufactured tetraethyl lead, promote it, while car manufacturers and environmental organizations oppose it.

The manufacturers say that manganese emissions interfere with the performance of the diagnostic computers they build into their cars, and that MMT causes more unburned hydrocarbons to be released through the car's exhaust. Environmental organizations are mainly concerned with the health consequences of using MMT. While manganese in milligram quantities is an essential human nutrient, inhaling manganese compounds can have serious consequences.

Researchers have linked a condition known as parkinsonism — a nervous disorder that resembles Parkinson's disease — to manganese. Steelworkers and welders, who are occupationally exposed to the metal, are more likely to be affected. When they get away from sources of manganese, those afflicted with parkinsonism improve. And now researchers are also looking into whether Parkinson's disease itself is linked to manganese exposure. For the time being, Canadian and U.S. law permits MMT use, but some gasoline companies have opted not to use it.

90. What momentous second-millennium achievement involved the chemical reaction between nitrogen tetroxide and monomethylhydrazine?

———

The launch of the lunar lander from the surface of the moon. Landing astronauts on the moon and returning them safely to Earth is one of humankind's greatest achievements.

In one sense, launching a rocket from the moon is easier than launching it from Earth. The moon has only about one-sixth the gravitational pull of our planet. Furthermore, since there is no air on the moon, air friction won't slow the rocket down after launch. That's why the lunar lander's engineers did not have to give their creation a sleek, aerodynamic shape, as they would have if the rocket was intended for an Earth launch. In fact, the thing looked more like a strange bug than a rocket.

The launch from the moon's surface required a foolproof fuel-oxidizer system. Since a rescue was out of the question, the system had to ignite with virtually no chance of failure. The engineers chose monomethylhydrazine as the fuel and nitrogen tetroxide as the agent that would provide the oxygen needed for combustion. We refer to this mixture as "hypergolic," meaning that the combination of the reagents results in instant ignition. There is no need to initiate the reaction with any other ignition system. As we know, the engineers got it right. The launch from the moon worked flawlessly on six occasions.

91. In 1803, a young German pharmacy assistant named Friedrich Serturner isolated a substance from a natural source. If used wisely, the substance would relieve untold suffering; but if it was used unwisely, it would *cause* untold suffering. What was it?

Morphine. Serturner was quite familiar with the medical use of opium, as any pharmacist of his day would have been. This exudate of the poppy had been in continuous use for some four thousand years — the ancient Sumerians described it in their writings.

The ancient Egyptians, Romans, and Greeks all knew of the pain-relieving properties of opium, and Shakespeare certainly knew about its sleep-inducing effect. He refers to it in *Othello*: "Not poppy, nor mandragora, / Nor all the drowsy syrups of the world, / Shall ever medicine thee to that sweet sleep / Which thou ow'dst yesterday." Indeed, Serturner named the specific component he isolated from opium morphine, after Morpheus, the Greek god of dreams.

By isolating morphine, Serturner made its medical application more exact, because now physicians could prescribe specific doses. The advent of the hypodermic syringe made morphine easy to use, although this increased its abuse potential. Opium had always been a popular euphoria-inducing substance, but most opium lovers ate or smoked the poppy extract. English Romantic writers made the stuff fashionable. In 1797, Coleridge wrote *Kubla Khan* and claimed he'd composed it in an opium-induced sleep; in 1922, Thomas de Quincey produced a remarkable work called *Confessions of an English Opium Eater*.

During the twentieth century, morphine abuse became a major social problem, and authorities moved to regulate the production and sale of the drug. Unfortunately, they failed to gain control of it, and morphine addiction remains a huge concern. Still, for patients suffering intense pain, morphine is a wonder drug. Perhaps more than any other substance, it demonstrates how a chemical can do good or bad, depending on how we use it.

92. Some European countries have enacted a law stating that jewelry cannot contain more than 0.05 percent of a certain substance by weight. What is the substance, and what prompted the law?

The substance is nickel, and legislators enacted the law because nickel is an allergen. The word *nickel* comes from the German for "devil," because the metal interfered with the smelting of copper; German miners in the 1700s called it *Kupfernickel*, or "copper devil."

Nickel really may be the devil to some people because of its allergenicity. Researchers in Finland have recently reported that body piercing is likely responsible for a dramatic increase in nickel allergies. Nickel-allergy sufferers can remain sensitive to the metal for life, and they also have a greater risk of developing other allergies. Many sensitized people have trouble finding wristwatches, belt buckles, or eyeglass frames that they can wear. The researchers discovered that even jewelry that tested negative for nickel by a standard color test involving ammonia and dimethylglyoxime released nickel when exposed to artificial sweat. Jewelry intended for pierced tongues, cheeks, and genitals fared the worst: eleven out of the twelve pieces sampled exceeded safety standards.

Some people are so sensitive to nickel that they even react when they handle coins. Allergic reactions to money are not restricted to coins. In Australia, there have been reports of people developing dermatitis on their hands from touching the banknotes in their pockets. Australia uses plastics called acrylates in its banknotes, and researchers have conducted patch tests that confirm a possible contact allergy to acrylates. Talk about money burning a hole in your pocket!

93. All books that deal with the history of chemistry refer to the philosopher's stone. What was it made of?

This is somewhat of a trick question — the philosopher's stone never existed. But try telling that to an alchemist.

Alchemists believed that certain metals ripened into gold in the earth. They based this belief on the fact that animals and plants grow from seemingly nothing into finished products. Gold, they thought, must surely form through a similar process. But, they asked themselves, what substance did the earth harbor that could facilitate such growth? It had to be the philosopher's stone.

They concocted all kinds of recipes for it. One of the most famous involved two thousand eggs and an Italian alchemist named Bernard Trevisan. In the fifteenth century, Trevisan embarked upon his alchemical quest by boiling the eggs, removing their shells, and separating the whites from the yolks. He proceeded to putrefy the whites with horse manure, then he combined the mess with the eggshells. Next, he heated the concoction and distilled from it an oil, which he allowed to harden into the "philosopher's stone."

Despite such efforts, the alchemists never found a way to turn metals into gold, but their philosopher's stone concept was an interesting one. Today, we would refer to it as a catalyst

— a substance that enhances chemical reactions without being consumed. So, we can regard modern catalysts as philosopher's stones. They may not turn metals into gold, but they turn oil into margarine and car-exhaust gases into carbon dioxide, nitrogen, and water.

94. During the 1800s, physicians carried walking sticks tightly wrapped with strips of leather. Why?

Pain relievers were scarce in those days. When a physician had to perform a particularly painful procedure, he jammed his walking stick sideways into the patient's mouth. The unfortunate soul would bite down on the stick and endeavor to cope with the pain. It was sort of like biting on a bullet.

But the cane symbolized the physician even before this rather inventive technique became widespread. In the seventeenth century, doctors carried canes with hollow gold heads full of garlic-laced Marseilles vinegar, which was rumored to have kept four plunderers of dead bodies from contracting the plague. To ward off infection, the doctors would repeatedly sniff the vinegar.

The idea of carrying a stick in the first place may have belonged to Paracelsus, the famous physician/alchemist who lived a couple of hundred years earlier. When his treatments failed, Paracelsus would sometimes dispense a mysterious powder from his hollow stick or sword handle, administering it to his patient with a flourish. This placebo treatment often had a miraculous effect, bolstering Paracelsus's theory that "Imagination takes precedence over all."

95. What do red clover, black cohosh, and soy have in common?

All are sold as treatments for menopausal symptoms. Somewhere between the ages of forty-five and fifty-five, women experience the onset of menopause. This change of life usually declares itself with hot flashes, a sign of diminishing estrogen levels.

Doctors advise many menopausal patients to counter their symptoms with hormone replacement therapy, but this approach isn't appropriate for all women. A family history of breast or uterine cancer precludes estrogen treatment, as does a history of chronic liver disease.

Hormone replacement therapy has a checkered history. At one time, the medical community was convinced that it reduced a woman's risk of heart disease after menopause, but the accumulated data don't bear this out. While there is evidence that the therapy increases bone strength and reduces the risk of osteoporosis, studies have also shown a link with breast cancer. These days, a significant number of women are looking instead to natural therapies, which they perceive to be safer.

Soy and red clover contain compounds called phytoestrogens (estrogens from plant sources), so it's not surprising that they can reduce menopausal symptoms. But they don't do it as effectively as estrogen does. Phytoestrogens have only one-tenth to one-one-hundredth the potency of estrogen in attaching to estrogen receptors in cells. For some women, this effect is sufficient, but many others are disappointed. For example, in one study, women taking seventy-six milligrams of soy isoflavones a day had 45 percent fewer hot flashes after three months. That sounds impressive until you learn that women given placebos experienced a 30-percent reduction.

Studies with red clover show that it works no better than placebo for hot flashes, although it may have some effect in reducing vaginal dryness. Black cohosh has been shown in some clinical studies to be effective in the short-term treatment of menopausal symptoms. Its mechanism of action is not clear because its supposed active ingredients, terpene glycosides, don't behave as estrogens. The interaction of these natural supplements with medications has not been well studied. Black cohosh, for example, may interact adversely with the breast cancer drug tamoxifen. Unlike prescription estrogens, the natural alternatives have not been shown to reduce the risk of osteoporosis.

There is another natural way to reduce hot flashes: work out. Women who exercise regularly are less likely to have them.

96. What is the connection between sand and the computer age?

Sand is composed of silicon dioxide, the most abundant compound in the Earth's crust. Heating sand with carbon to a temperature of about 2,200°C (3,992°F) can convert the silicon dioxide to carbon monoxide and elemental silicon. This reaction made the computer age possible.

Silicon is a semiconductor, meaning that it conducts electricity under some conditions but not others. This property is critical to the production of transistors and integrated circuits. And there's another connection. The glass that shields our computer screens is also made of sand.

Glass is manufactured by heating sand and allowing it to cool down. The silicon and oxygen atoms of silicon dioxide have a highly ordered arrangement — in other words, silicon dioxide has a well-defined crystal structure. When sand is liquefied by heat and the liquid is then cooled, this orderly arrangement

of atoms is lost and the more random pattern characteristic of glass appears. The sand has been "vitrified." Pure sand has a very high melting point and becomes viscous when it melts. But, if sodium oxide (soda) and calcium oxide (lime) are mixed in to the extent of fifteen percent and ten percent respectively, both the melting point and viscosity are reduced. The resulting "soda lime" glass is what we use for windows, bottles, and computer screens.

Since glass expands when heated and contracts when cooled, rapid cooling can make it crack. If we add boron oxide to the glass mix, we greatly reduce the expansion and contraction. This is how we produce Pyrex, or laboratory glassware. We can alter the way glass reflects light by mixing in some lead oxide, creating heavy crystal ware. Colors in glass are the result of impurities — such as cobalt oxide (blue), chromium oxide (green), and iron sulfide (amber).

97. What do you get when you mix wood ash with tristearin?

Soap. According to Pliny, the Phoenicians made the first soap about 2,500 years ago from boiled goat fat and caustic wood ash. The recipe hasn't changed much: we still manufacture soap by boiling some sort of fat with a base. The most common base is sodium hydroxide, or lye, hence the term "lye soap" for old-fashioned caustic soaps. Remember Granny on *The Beverly Hillbillies* brewing up batches of lye soap out by the "cee-ment pond"? Probably not.

We don't really know who discovered this reaction. Historians tell us that Roman women noted the remarkable cleaning properties of clay taken from the banks of the Tiber River at the foot of Mount Sapo. People burned sacrificial offerings to the gods on the mount, and the animal fat reacted with the hot wood ash to form soap; rain washed this substance downhill to the river, where the clay absorbed it. What we do know for sure is that soap making was an established enterprise in Europe by the seventh century, and commercial soap manufacturers were in business in England by the twelfth century. But their products were expensive, and to most people they were an unobtainable luxury. Of course, expense was no obstacle to Queen Elizabeth — she had a bath every three months, whether she needed it or not.

Tristearin is simply the chemical term for a type of fat. Fats are composed of fatty acids joined to a molecule of glycerol, so they are also called triglycerides. Tristearin, a fat found in tallow, is made of three molecules of stearic acid linked to a glycerol molecule. When it is chemically reacted with a base — like sodium carbonate, which is found in ashes — it forms sodium stearate, otherwise known as soap.

Soap molecules have a water-soluble end and a fat-soluble

end. They clean by anchoring the fat-soluble end in oily dirt, and the dirt is then removed by rinsing with water, to which the other end of the molecule is attracted.

98. The risk of vitamin B12 deficiency increases as we get older. Why?

As we age, our stomachs produce less acid. Vitamin B12 in food is bound to proteins, which the body must break down before it can absorb the vitamin. It needs stomach acid to do this.

Signs of vitamin B12 deficiency can take a long time to show up, and they can be subtle. We require B12 in order to maintain myelin, the coating that protects our nerve cells. Should our myelin wear away, we may experience tingling sensations, numbness, and a burning feeling in the limbs, as well as memory problems and mood disorders. A more dramatic effect of vitamin B12 deficiency is impairment of the red blood cells' ability to carry oxygen. In the absence of B12, these cells grow abnormally large as they try to capture more oxygen. This is called macrocytic anemia.

The general dosage recommendation for people over fifty is a B12 supplement containing at least the daily value of 2.4 micrograms, but many scientists recommend higher doses — up to twenty-five micrograms. There is no risk of overdosing.

A Tufts University study on B12 yielded a surprising result. Researchers tested three thousand healthy young adults with healthy diets and found that 40 percent had low blood levels of B12. It seems that our bodies do not absorb B12 from meat, chicken, and fish as well as we thought. Perhaps cooking binds the vitamin more strongly to proteins. We absorb the vitamin from dairy products much better. A glass of milk can give us a microgram of B12.

So, what's the bottom line? We should all take a daily supplement of vitamin B12 and ensure that our diets include enough dairy products. And we should do it while we can still remember to — B12 deficiency impairs memory.

99. What common metal was once more valuable than gold?

Aluminum. If you had been invited to dine at the court of Emperor Louis Napoleon III in the 1800s, you might have noticed that the table was laid with gold cutlery. That's because you weren't important enough to merit the good stuff — aluminum! At the time, this metal was more precious than gold. It was so highly valued that a small ingot of it was displayed next to the crown jewels at the Paris Exposition of 1855.

Today, this seems unimaginable. Aluminum is everywhere. We drink from aluminum cans, fly in aluminum airplanes, play aluminum-coated CDs, wrap our food in aluminum foil, and, yes, eat with aluminum utensils. What has changed? Certainly not the availability of aluminum — it has always been the most abundant metal in the Earth's crust. The problem is, it's found in combination with other elements — such as oxygen and silicon, as in clay and rocks.

A good example is bauxite, an ore named after Les Baux, in France, where it is found. But converting bauxite into pure aluminum is quite an involved process. Bauxite's reddish color is due mostly to iron compounds, which must be removed in order to obtain pure alumina, the white powder that is the key to isolating aluminum. No commercially viable method to convert bauxite to aluminum existed until the late 1800s, when Charles Martin Hall in America and Paul Louis Toussaint Héroult in France independently observed the same phenomenon. Each discovered that passing an electric current through a solution of pure aluminum oxide in cryolite (sodium aluminum fluoride) produced the pure metal. Both men were only twenty-two years old at the time.

Modern processing is based on the Hall-Héroult method, which makes inexpensive aluminum widely available. Aluminum is a remarkable metal. It does not corrode the way iron does, and we can recycle it easily — aluminum cans are continuously melted down and reused. One of the metal's most important properties is its low density. It's very light. That's why the Soviets used it to construct Sputnik, the world's first satellite. And the space shuttle is made primarily of aluminum as well. The huge fuel tank that dominates the vehicle is an aluminum alloy. Aluminum even gets the shuttle into outer space — it's the fuel used in the solid fuel boosters. That's because aluminum burns in the presence of an oxidizing agent, providing the thrust needed to get the shuttle into orbit so that it can meet up with the space station.

On a clear night, look skyward and try to catch a glimpse of the space station as it orbits the Earth. It will be very shiny. Do you know why? It's made mostly of aluminum!

100. An epidemic is a disease that attacks many people in the same region at the same time, spreading rapidly. A widespread epidemic is called a pandemic. What was the worst pandemic in human history?

———

It wasn't the bubonic plague, typhoid, smallpox, or AIDS. It was the flu pandemic of 1918.

The pandemic began on U.S. military bases. Within six weeks, 3 percent of the recruits at Camp Sherman died. Troops carried it to Europe, and from there it spread as far as China. Within a year, twenty-five million people died — a staggering number when compared with the World War I death toll of nine million. And the ships that crossed the Atlantic arrived with far fewer passengers than they departed with.

The pandemic had a powerful economic effect as well. Sales of flu remedies, none of which was effective against the virus, skyrocketed. As it turned out, the product that may have benefited most from the scourge was Vicks VapoRub, introduced in 1905 by Lunsford Richardson, a druggist in Selma, North Carolina. The rub's main ingredient was menthol, a compound extracted from oil of peppermint.

Richardson was not the first person to realize menthol's potential in pharmaceutical products. That honor goes to Jules Bengue, the French pharmacist who had created Ben-Gay as a treatment for sore muscles. Bengue had noted that menthol produced heat when you rubbed it on your skin. He combined it with the painkiller methyl salicylate and sold the concoction as a treatment for arthritis, gout, and neuralgia. Some Ben-Gay users with colds reported that sniffing the product cleared their sinuses.

Richardson sold Ben-Gay, and he heard such testimonials from his customers. So he blended menthol into petroleum jelly and produced Richardson's Croup and Pneumonia Cure Salve. When rubbed on the chest, the stuff did produce the sensation of heat, and as the menthol evaporated it did clear the nasal passages. The product's name was too long, however, so Richardson sought a shorter, catchier one. He had a physician brother-in-law called Joshua Vick, and Vick had given him access to the laboratory where he'd created his salve. A grateful Richardson decided to name it Vicks VapoRub, and the name stuck.

101. The Calabar bean grows on a woody climber plant, much like the garden pea or runner bean. Another name for it is ordeal bean. Why?

———

Some primitive societies used the bean, which grows in Africa, to determine whether an accused person was guilty of a crime. Interrogators forced the accused to eat some Calabar beans. If he was guilty, he died. If he was innocent, he vomited and purged himself of all blame. Or so they claimed.

Today, we know that the Calabar bean contains physostigmine, a very dangerous substance. It has a dramatic effect on the human nervous system because it enhances the activity of one of the body's most important neurotransmitters, acetylcholine. An excess of physostigmine leads to overstimulation and death. Physostigmine is sometimes used in modern medicine to counteract poisons that work by impairing the activity of acetylcholine. It has also been used to treat glaucoma, a condition in which pressure builds in the eyeball. But one wonders how many innocent people lost their lives to the "ordeal" of the bean, and how many guilty people were saved by their strong stomachs.

102. During World War II, the U.S. military had to confront a serious medical problem. In a desperate attempt to find a solution, prison doctors in Chicago infected about four hundred inmates with what illness?

———

Malaria. The Americans entered Guadalcanal in June 1942, and by December more than 8,500 U.S. soldiers had landed in hospital with malaria. Quinine was an effective treatment for malaria, but the U.S. did not have enough of the drug in reserve.

The only available drug was quinacrine, a synthetic analogue of quinine, which scientists had synthesized in 1934.

Atabrine, as this drug was called, had a variety of side effects. Those who took it often experienced headache, nausea, and vomiting, and their skin would turn a sickly yellow hue. Rumors started circulating that Atabrine caused impotence. But Atabrine did work against malaria, and, in 1944 alone, drug companies churned out 3.5 billion tablets. Desperate to produce more quinine, a better drug, the U.S. government mounted the Cinchona Mission in 1942, sending teams of scientists to South America to find quinine-rich cinchona trees, which could produce copious amounts of quinine.

In a parallel effort, prison doctors in Chicago infected about four hundred (probably unsuspecting) inmates with malaria in order to test potential cures. The undertaking was so nefarious that, during the Nuremberg trials, Nazi doctors actually referred to it to justify their own experiments on prisoners.

Finally, in 1944, Robert Woodward synthesized quinine from coal-tar derivatives, but the synthesis was not commercially viable. Chloroquine, a close chemical relative of quinine, did yield to synthesis and remains the mainstay of malaria treatment.

103. Who gave the following advice about weight loss? "Those desiring to lose weight should perform hard work before food. They should take their meals after exertion and while still panting from fatigue. They should, moreover, eat only once a day and take no baths and sleep on a hard bed and walk naked as long as possible."

Hippocrates, the father of medicine. While nobody has conducted research on the effects of running naked through the streets (for which the ancient Greeks had a propensity, as we know from Archimedes's naked romp), it's interesting that Hippocrates recognized the relationship between food intake and energy expenditure some 2,400 years ago. Perhaps scientists should examine Hippocrates's ideas, because in America today 55 percent of adults and 10 percent of children are overweight.

104. Why shouldn't we refreeze previously thawed food?

There are two issues involved: bacterial contamination and the deterioration of texture and flavor. Let's deal with the contamination issue first.

Bacteria like to eat the same kinds of things we do, so they get into a lot of our food. They also need water to survive, so they find moist environments particularly hospitable. Dry foods, such as salt, sugar, and rice, are rarely contaminated by bacteria. Just because there is bacteria in a given food, however, doesn't mean that we'll get sick if we eat it. Much depends on the type of bacteria and — most important — the quantity. It usually takes a large number of bacteria to cause illness.

Freezing does not kill the bacteria, but it greatly slows their multiplication rate. When frozen food thaws, bacterial growth speeds up, especially if we allow the food to reach room temperature. If we then decide to refreeze the food, there will be more bacteria than there were in the first place when we thaw the food again, and they'll keep on multiplying. Some people wonder why this should matter. After all, when we cook the food, the bacteria will be destroyed. Heat indeed kills many bacteria, but some varieties — staphylococcus, for one — produce a heat-stable toxin.

So, here are the appropriate guidelines. You can safely re-freeze food that still has ice crystals or that has been no warmer than 5°C (40°F) and has been out of the freezer for no more than twenty-four hours. Do not refreeze shellfish, dishes containing cream, or cooked foods, because they are particularly prone to bacterial growth. You can often tell if a frozen food has previously been thawed: its package will be frost-coated, because when the food thawed moisture escaped from it and then refroze.

The "Do not refreeze" instruction originally came from Clarence Birdseye, the father of the frozen-vegetable industry. He wasn't worried about safety. He just didn't want people to remove his products from the freezer and then put them back, because this would cause textural damage and reduce sales.

Freezing does diminish the quality of food, mostly by affecting moisture content. It is well known that water can pass through cell membranes, as dictated by the principle of osmosis. Basically, water flows from an area of low concentration to one of high concentration. Since dissolved solids are more concentrated inside a cell, water flows in. But freezing undermines this osmotic capability and water begins to seep out of cells. This means that the pressure caused by the swelling of one cell against another is reduced and textural rigidity is lost. Also, when ice crystals form, they take up more space than the corresponding amount of water, and this expansion bursts cell walls and pushes cells apart, giving the released water an easy escape route. The result is a soggy mess.

The higher a food's water content, the mushier the food will be after freezing and thawing. Compare a thawed steak and a refrigerated steak. The difference in texture will be obvious. When liquid flows out of ruptured cells, it carries with it some flavor and nutrients, diminishing the food's quality. Vegetables survive freezing better than meat, seafood, and fruit.

Cryonics researchers interested in preserving human bodies in a frozen state (with the aim of "reanimating" them at some future point) must address this texture problem. We currently have no way of eliminating cellular damage caused by freezing, but these researchers are looking to change that. One of their avenues of pursuit is to examine certain proteins present in the blood of fish that live in freezing oceans. If you were frozen for a few hundred years and they managed to find a way to bring you back to life, you wouldn't want to wake up limp and oozing water. And it wouldn't be safe to refreeze you.

Incidentally, you can keep frozen hamburger for about three months. Steaks will last a year; chicken, nine months; and vegetables, eight months. The general rule is don't keep anything frozen for more than a year — not even people.

105. What fascinating method of closing wounds predates sutures and makes use of a living species?

Here is the method: pull the sides of the wound together; take a red ant and position it in such a way that its bite holds the wound closed; immediately cut off the ant's body, leaving its head in place as a clamp. Repeat with more ants until the entire wound is sealed. This technique was used as late as 1912, during the war between the Greeks and the Turks, and there are reports that Honduran tribesmen still implement it.

106. In 1898, as Pierre and Marie Curie were investigating an ore called pitchblende, they discovered that, even after they had removed its uranium content, the ore remained radioactive. They isolated the source of the radioactivity, which turned out to be a new element — a brilliant white metal that glowed with a faint blue light. What was it?

———

Radium. The Curies named the new element from the Latin word *radius*, meaning "ray," because of the rays it emitted in the dark.

At the time, no one knew about the dangers of radioactivity, and watch and clock manufacturers started using the substance to illuminate dials so that they could be read in the dark. Tragically, many of the women employed to paint the dials with luminous paint ended up with cancers of the mouth and tongue from licking their paintbrushes as they worked.

Quacks also got into the radium game. During the 1920s, a health drink called Radithor was sold throughout the U.S., but at a dollar a bottle only the rich could afford it. It was the brainchild of William J. A. Bailey, an entrepreneur who specialized in impotence cures. "Improved blood supply sent to the pelvic organs and tonic effects upon the nervous system generally result in a great improvement in the sex organs," claimed the ads. One of Bailey's best customers was Eben MacBurney Byers, racehorse owner, golf champion, and noted Casanova. Byers drank several bottles daily, and he sent bottles to his girlfriends. But soon his teeth started to fall out. Then his jaw disintegrated. Holes formed in his skull. He was soon dead of cancer. Byers's death killed the so-called mild radium therapy movement. In 1949, twenty years after Bailey's death, researchers exhumed his body and found that it was still radioactive.

107. How do they make instant coffee?

Manufacturers put roasted coffee beans in gigantic columns — up to two thousand pounds of beans at a time — through which they pass hot water or steam. Coffee drips out the bottom, and it evaporates to form a concentrate known as coffee liquor.

This liquor is then dried using one of two methods. Method one involves pouring the liquor through a huge, cylindrical drier, which is heated to a high temperature. This results in a dry powder dropping out the bottom. Method two involves freezing the coffee liquor into blocks, breaking the blocks into granules, and putting the granules in a vacuum chamber, where the water evaporates almost instantly. The latter technique generally results in a better flavor — it doesn't rely on high temperatures, which destroy some of the flavorful components.

Freshly brewed coffee contains about seven hundred compounds, and many of these contribute to the beverage's taste and smell. But when the hot water or steam passes through the coffee column, some of these compounds are lost because they are volatile. Manufacturers must recover them and add them to their instant coffee powder. Here's how they do it. The gases that evolve during the brewing process are cooled with liquid nitrogen until they solidify and are then dissolved in coffee oil under pressure. Just before the coffee is packed, this flavored oil is mixed into the coffee powder.

Manufacturers remove some undesirable compounds, such as methyl mercaptan and acetaldehyde, which give coffee an off-taste, from the coffee vapors by passing these through a filter made of zeolite, a type of clay that has an affinity for the compounds. Almost presto, we have instant coffee!

108. How can you get a peeled, hard-boiled egg inside a bottle with a neck smaller than the egg's diameter?

Drop a burning match into the bottle, wait a few seconds, and put the egg on top. The bottle will suck it right in! The heat of the match causes the air in the bottle to expand, forcing some air out of the bottle. As the air inside cools, a partial vacuum forms because there is now less air in the bottle. The air pressure outside is greater than it is inside, so the egg is essentially pushed into the bottle. You can remove the egg by turning the bottle upside-down, so the egg falls into the neck, then blowing vigorously into the bottle. Now the pressure inside becomes greater, and the egg pops out.

There's a common misconception that the partial vacuum needed to accomplish this trick occurs because oxygen is used up as the match burns. Oxygen is indeed used up, but it's replaced by the products of combustion: carbon dioxide and water vapor. It is the cooling of the warm gases that reduces the pressure.

109. Constantin Fahlberg, a German researcher, worked under Ira Remsen, one of the era's most famous American chemists, at Johns Hopkins University. One day, in 1879, Fahlberg ate a piece of bread in the lab. He noticed that it had a remarkable taste — a taste that would launch an industry. What had Fahlberg discovered?

Saccharin. The taste that Fahlberg noticed was sweet, and he traced it to a compound he had just been handling. Envisioning this substance's potential, he developed a commercial synthesis, and in 1885 he took out a patent.

Fahlberg called his discovery saccharin after the Latin *sac-charum*, for "sugar." The industrial production of saccharin got under way in 1900, and its popularity soared during World War I, when there were sugar shortages. Eventually, though, more and more people began to use saccharin because they recognized it as a calorie-cutting aid.

Saccharin is hundreds of times sweeter than sugar, and we don't absorb it for energy, so it's noncaloric. In the U.S., saccharin is allowed both as a tabletop sweetener and as a food additive, while in Canada it can only be sold as a tabletop sweetener. Safety questions about saccharin, which have now been resolved, date back to 1977, when researchers gave pregnant guinea pigs the saccharin equivalent of 1,200 to 1,800 soft drinks a day. The guinea pigs showed no ill effects, but, when the researchers fed their offspring the same dose, half the males developed bladder cancer. What damaged their bladder cells was likely the amount of saccharin administered, not a specific chemical activity. Large amounts of sugar and salt can cause similar damage. Diabetics have consumed saccharin for a hundred years with no ill effects.

Accordingly, in December 1998, a U.S. advisory committee, the National Toxicology Program, voted to grant saccharin a clean bill of health and withdrew the requirement that manufacturers put a warning label on saccharin packets declaring that the substance caused cancer in rats.

110. What is in windshield washer fluid?

Windshield washer fluid is a 40-to-60-percent mix of methanol and water. It contains a little detergent to remove dirt, and manufacturers also add a blue dye, because market studies have shown that people associate the color blue with being clean.

The reason we use the fluid instead of water is that it freezes at a lower temperature. Dissolving any substance in water reduces the water's freezing point. Interestingly, it does not matter what we add. What matters is how much we add: it is the number of foreign molecules that determines the extent of the freezing-point reduction. Essentially, the solute molecules get between water molecules and prevent them from joining together to form crystals. Sugar or salt would do the job, but they would leave a residue on the windshield as the water evaporated.

So, we need a volatile liquid that mixes easily with water. Methanol is ideal. A 40-percent solution in water will lower water's freezing point to approximately minus forty degrees, which is usually enough. The English chemist Robert Boyle first made methanol in 1661, when he heated wood in the absence of air. The wood didn't burn, but it decomposed to yield a variety of substances, one of which he named methyl alcohol, or methanol. The term comes from the Greek *methe*, for "wine," and *hyle*, for "wood," because methyl alcohol was considered to be the "wine of wood."

Until about 1930, Boyle's methanol-making method remained in use. Today, it is made from "synthesis gas." When steam reacts with coal, oil, or natural gas, it forms a mix of hydrogen and carbon monoxide. This mixture is known as "syngas," and it can be reacted with a zinc-oxide/chromium-oxide catalyst to make methanol. Methanol is used as a gasoline additive (it makes the gas burn better), as an industrial solvent, and, of course, as an ingredient in windshield washer fluid.

In a pinch, you can also use a homogeneous mixture of water and ethanol as windshield washer fluid — just fill the container in your car with vodka.

111. Five soldiers in the former Soviet Union were building airplane models of Plexiglas. As they glued the pieces together with a solvent called dichloroethane, one of them noticed that it smelled sweet. They decided to see if they could get high on it. The men poured some into a metal drinking cup, diluted it with water, and passed the cup around. The first soldier got mildly ill, the others became ill progressively, and the last one died. What happened?

Dichloroethane dissolves Plexiglas — that's why we can use it as a glue. When the softened surfaces are pressed together, they begin to stick as the solvent evaporates. Dichloroethane does have a sweet smell, and the soldiers had undoubtedly heard about people getting high by sniffing or drinking solvents. But they didn't know how toxic this particular compound is.

Neither did they know that it doesn't mix with water. Dichloroethane is denser than water, and it sank to the bottom of the soldiers' metal cup. So, the first man drank mostly water. By the time the cup made it to the last man, the water was gone. He drank undiluted dichloroethane and died.

The tragedy might have been averted if the men had used a glass instead of a metal cup. This would have revealed the separate layers. But then they might have decided to drink the stuff straight, in which case they would all have perished.

112. We routinely tell students not to wear false fingernails in the chemistry lab. Why?

False fingernails are made of a plastic, poly(acrylonitrile-butadiene-styrene), which has roughly the same flexibility and stiffness as real fingernails. They are coated with polyacrylate, which gives them shine, and the wearer attaches them to her real nails with a cyanoacrylate glue, which is essentially Krazy Glue. Unfortunately, false fingernails are flammable.

When subjected to the heat of a Bunsen burner — 500°C (932°F) — false nails can ignite in less than a second, becoming fiery balls of molten plastic that adhere to the fingers. But you don't need a Bunsen burner to ignite false nails — a candle will do it. At a temperature of 300°C (572°F), it takes slightly longer, but the false nails still catch fire in a second or a second and a half. So, the moral is don't play with, or near, fire if you're wearing false fingernails. And the longer the nails, the greater the danger.

113. What is the strongest existing fiber, synthetic or natural?

Believe it or not, it's spider silk, the stuff spiders use to weave their webs and catch their prey.

Spiders also use silk to hang from a support and swing back and forth as they build their webs. Their silk fibers are at least four times as strong as steel fibers of the same diameter, stronger even than fibers of Kevlar, the synthetic material used to make bulletproof vests. Not only is spider silk strong, but it's also stretchy. It has to be, otherwise flying insects would just bounce off it.

The properties of spider silk reflect its internal molecular structure. All such silk is made of protein, but spiders produce a variety of proteins with which to spin it. Proteins are built of amino acids, but the specific properties of a protein are determined by the variety of amino acids that join together to form the protein chain. Scientists have found six amino acids — namely, glycine, alanine, proline, tyrosine, serine, and glutamine — in spider silk. Chains that have long sequences of alanine units can pack together very closely in what we call beta-sheets, and they are responsible for the strength of the fiber. Flexible chains have sequences of amino acids that allow them to fold and act like miniature springs.

Amazingly, spiders can customize their silk. They have several silk glands, and each has a different secretion; they can blend the various proteins that their glands produce to create a silk that's suited to a particular purpose. Unfortunately, we cannot raise spiders in captivity to produce silk. Aggressive creatures, they will attack and eat each other. Bacteria, by comparison, are more peaceful. And they can be goaded into making spider silk.

The DNA sequences that make up the spider genes that code for the silk protein have now been identified. When these genes are inserted into bacteria, they produce spider silk — but not in industrial quantities, and that's the problem. In an attempt to solve this problem, Montreal's Nexia Biotechnologies implanted spider silk genes into goats, and the proteins needed to make the silk showed up in the goats' milk. Eventually, this research may allow us to fabricate superior bulletproof vests and superstrong fabrics and cables, not to mention artificial tendons and ligaments.

114. Bernard Courtois owned a factory that produced saltpeter, or potassium nitrate, an essential component of gunpowder. This involved burning seaweed, a source of potassium. One day in 1811, as his workers cleaned one of the seaweed-burning tanks with a strong acid, the room filled with violet fumes. What were these fumes?

Iodine. This element is found in seawater and seaweed in the form of iodide ions, which can be oxidized to iodine with sulfuric acid.

After the fumes had condensed on the cold surface of the tank, Courtois noticed a residue of dark purple, metallic-looking crystals. A sample of this made its way into the hands of French chemist Joseph Gay-Lussac. He'd heard that Sir Humphrey Davy also had a sample, and he didn't want an Englishman to make a big discovery based on a Frenchman's shrewd observation, so he got to work. Gay-Lussac experimented feverishly and discovered that the purple material was a new element. He named it *iode*, from the Greek for "violet." Davy suggested "iodine," echoing "chlorine," an element that iodine resembles.

An application for the new element was soon found. In 1820, Geneva doctor Jean-François Coindet guessed that it was the stuff that is present in the ashes of sponges, which physicians used to treat goiter. Today, iodide is used to treat some forms of goiter because it supplies the raw material needed to produce thyroxine, an essential thyroid hormone. Goiter is rarely seen in people living near the sea, because seafood is rich in iodine. In order to prevent goiter in the landlocked, 0.01 percent potassium iodide is commonly added to salt.

Iodine has antiseptic properties and can therefore be used to disinfect cuts as well as drinking water. Silver iodide is used to make photographic film and to seed clouds for rain — it forms enormous numbers of tiny crystals, which act as nuclei for raindrop formation. Physicians use radioactive iodine, I-131, to diagnose thyroid disease, measuring the rate at which the gland takes up the iodine. At a higher dose, radioactive iodine, since it concentrates in the thyroid, can be used to treat thyroid cancer without damaging other organs.

115. A three-story, yellow-brick house on a London, Ontario, street corner has been turned into a museum. About three thousand people visit it annually. Many of them are physicians, and they head for the bedroom. There they sit on the bed and reflect on one of the greatest ideas in medical research, which was supposedly hatched in that bed. Whose bed was it?

———

Frederick Banting. Dr. Banting moved to London at the age of twenty-eight to set up a medical practice. He took on a second job as lecturer in the Department of Surgery and Physiology at the University of Western Ontario; for this he was paid two dollars per hour. On the evening of October 30, 1920, Banting was preparing a lecture on carbohydrate metabolism and the pancreas when he came across a journal article about the relationship between the islets of Langerhans — which are clusters of cells in the pancreas — and diabetes.

Banting didn't sleep well that night because he was worried about his debts, his relationship with his fiancée, and the next morning's lecture. He got up at two o'clock in the morning and

wrote up some notes proposing an experiment that involved tying off a dog's pancreatic ducts to isolate the internal secretions of the islets. That night, the fate of diabetics all over the world changed. By 1921, scientists had extracted insulin, and by 1923, the drug company Eli Lilly was selling it under the name of Iletin.

Fredirick Banting died in an airplane crash in 1941 after convincing authorities to allow him to travel to England aboard a bomber. Soon after take-off, the plane began to have technical difficulties, and the pilot attempted an emergency landing in a remote area of Newfoundland. Banting survived the crash, but he was severely injured. It took days for rescuers to reach the crash site, and by then it was too late. They found Banting dead in the snow, a few feet from the wreckage.

116. Why must chlorine bleach never be mixed with ammonia?

Most people know — at least they should know — that mixing chlorine bleach with any acid, even vinegar, can produce dangerous amounts of chlorine gas. Mixing bleach with ammonia doesn't release chlorine, but it does form another dangerous compound: chloramine. While chloramine is slightly less damaging than chlorine, it is still hazardous. Fumes can cause immediate watering and burning of the eyes, runny nose, sore throat, coughing, shortness of breath, and breathing difficulties. Inhaling too much chlorine or chloramine can lead to chemical pneumonia.

Chloramine, however, has another side. It can disinfect our drinking water, so it serves as an alternative to chlorination. Chlorination is one of the all-time greatest technological innovations — before it appeared, cholera, typhoid fever, and dysentery were common. Yet chlorination does raise certain concerns. Naturally occurring organic compounds in water, as well as some pollutants, react with chlorine to form the notorious trihalomethanes, or THMs, which are likely carcinogens. Chloramines do not do this. That's why many water treatment authorities are switching to chloramine.

Chloramine is more stable than chlorine, and it lasts longer in the distribution system, providing increased protection. At the levels used in drinking water, about one part per million, chloramines are safe. At higher concentrations, the story changes. Anyone whose eyes have burned after swimming in a pool has experienced chloramine irritation. Many assume that chlorine is the culprit, but they're wrong. Chloramines form when sweat or urine, both of which release ammonia, combine with chlorine. So, if your eyes burn while swimming, chances are that someone has been fouling the water.

117. What poison did Shakespeare use to dispense with Romeo?

———

The Bard does not actually mention the poison by name, but he likely had in mind a certain substance well known to cause almost instant death. It was a preparation made from the *Aconitum napellus* plant, commonly known as monkshood or wolfsbane.

The active ingredient is aconite, and it's found mostly in the root, which resembles a turnip. It is such a powerful poison that a single milligram is enough to kill. The ancient Greeks knew of its potency — on the island of Ceos, infirm men who were no longer useful to the state were condemned to drink a brew made from it. In India, people poisoned wells with aconite so that the advancing British Army couldn't drink from them. Accidental poisonings have also occurred. In California, a twenty-three-year-old man on a nature walk north of Santa Cruz mistook monkshood for watercress, ate two plants, and died within fifteen minutes.

In England, in the 1500s, justice officials experimented with giving aconite to condemned criminals. They did this on two occasions, and both of their subjects died. It's very likely that Shakespeare knew about this. The identity of the sleeping potion that his Juliet takes to enter a state resembling death remains a mystery, however. But a well-measured dose of opium could do the job.

118. In 1856, during his Easter break from the Royal College of Chemistry, eighteen-year-old William Henry Perkin tried to make quinine in his improvised home laboratory. He failed, but the unsuccessful experiment brought him riches and fame. What had Perkin accidentally discovered?

———

The first synthetic dye. Failing to make quinine, Perkin instead brought about the foundation of the synthetic dye industry.

Perkin's professor, August Wilhelm Hofmann, had declared to his students that the first chemist to synthesize quinine would become wealthy and famous. The world desperately needed quinine to treat malaria, but the only source of it was the cinchona tree, which grew in South America. Hofmann had worked on substances isolated from petroleum, and he'd found that the chemical composition of some of these compounds was like that of quinine, except for the fact that they lacked oxygen. It therefore occurred to him that, by treating coal-tar extracts with chemicals called oxidizing agents, he could add the required oxygen to the molecule.

Perkin, eager to give the experiment a go during his school holiday, reacted the coal-tar derivative aniline with oxidizing agents. All he got was an ugly sludge. He had trouble cleaning the sludge off his glassware, so he tried dissolving it with alcohol. When he poured in the alcohol, the solution turned a beautiful purplish color, which he eventually called mauve.

Recognizing the importance of his discovery, Perkin went into business producing the mauve dye with his father and brother. Mauve became the first commercially successful coal-tar dye, and it brought riches and fame to its discoverer. In a sense, quinine did reward Perkin, but not exactly the way Hofmann had predicted.

119. During a four-year period, the hospitalizations of 183 American children under the age of seven and the deaths of five others were attributed to television. In what way was television responsible?

These tragedies had nothing to do with electromagnetic radiation, exploding sets, or TV dinners. They were caused by toppling TVs. A report in the *Archives of Pediatrics and Adolescent Medicine* describes the study, conducted at the New England Medical Center in Boston, that identified the problem. In this age of big-screen TVs, television sets are proportionately heavier at the front and therefore less stable; simple design changes could reduce the toppling-TV problem. The number of people who sustain mental injuries from watching too much TV remains unknown.

120. In Leeds, England, children once amused themselves by throwing lit matches into the canal. But they achieved the effect they were looking for only on very cold mornings. What was the effect, and why didn't it occur on other occasions?

In Leeds, the outflow of sewage into the main canal was such that the basin became a virtual fermentation tank. Bacteria broke down the sewage, generating copious amounts of methane gas. Methane is highly flammable, and when it's ignited it flames spectacularly.

Canal bargemen worried about this phenomenon, but it delighted the mischievous children. Methane is lighter than air, and it dissipates quickly into the air. But its density depends on temperature, and on very cold days there was enough methane just above the surface of the Leeds canal to cause the flaming effect.

121. What were the ancient Romans talking about when they described orbicularis oris activity as *osculum, basium,* or *saviolum*?

Kissing. Orbicularis muscles control the lips, and for the Romans *osculum* was a peck on the cheek; *basium* was a lip-to-lip juxtaposition; and *saviolum* was a passionate kiss that involved the tongue.

Where did this strange habit come from? It seems to be cultural, not innate. People on South Pacific islands were great lovers, but they knew nothing of kissing before the Europeans arrived. Not all Europeans approved of kissing. By the late Middle Ages, the Catholic Church had decreed that kissing in reverence of God was acceptable, but kissing with the intent to fornicate was a mortal sin.

Scientifically, though, there may be a purpose for the kiss. The mouth and tongue are full of nerve endings, which, when stimulated, trigger the release of hormones and endorphins, lifting the mood of the kissers. But kissing may have another role. The original "Eskimo kiss" may simply be a technique to sample the odor produced by scent glands on the cheek. Studies show that women subconsciously prefer the smell of men who have genes that code for different immune system proteins than their own. Mixing immune system proteins can result in healthier offspring.

So, there may be something positive going on when saliva, sebum, bits of food, and about 278 different types of bacteria are exchanged.

122. When adding pineapple to Jell-O, why can we use the canned variety but not the fresh?

Jell-O is made of collagen, a protein found in skin and bones. When mixed with water, collagen forms a three-dimensional set of interlinked molecules, which act as a scaffolding to support water molecules. The result is a gel. Anything that breaks the protein molecules down will destroy this gel.

Pineapple contains bromelain — a proteolytic, or protein-splitting, enzyme. Enzymes themselves are specialized protein molecules produced by living systems, and they act as biological catalysts that speed up the numerous reactions that together constitute life. The bromelain in pineapple serves as a defense against predators. It irritates the tissues in the mouth as proteins begin to break down. Bromelain, like all enzymes, is heat-sensitive, and cooking destroys it. Canned pineapple is

heat-treated before it's canned, so it contains no bromelain, but fresh pineapple releases the enzyme, which breaks down the structure of the gelatin.

123. What recent French lifestyle trend has potential negative health implications?

A decrease in wine consumption. The French have a lower rate of heart disease than North Americans, despite the fact that they consume more fat, have higher cholesterol levels, and smoke more. Scientists have suggested that they are protected from heart disease by antioxidants in grapes, particularly red ones. Resveratrol, a compound found in the skin of red grapes, prevents cholesterol in the blood from converting to its oxidized form, which damages arteries.

But now, the French are drinking less wine. These days, the average Frenchman drinks about 60 liters of wine a year, down from 108 liters in 1960. The nation's young people are drinking less wine, and sales of soft drinks and bottled waters are on the rise. About 40 percent of the population say they drink wine just once a week; in 1980, this percentage was only 30.

Wine merchants say that they have noticed the drop, but they've also noticed that, while the demand for cheap table wines has ebbed, the demand for better-quality wines has increased. We will have to wait and see what this trend does to heart disease rates. *Sacre bleu!*

124. There has been a lot of talk about mercury in fish. How does mercury get into fish in the first place?

The Romans named this element after Mercury, the messenger of the gods. Its symbol, Hg, comes from the Latin *hydrargyrum*, meaning "liquid silver." Both of these derivations refer to mercury's mobility, and it is indeed a mobile element in more ways than one.

When liquid mercury is struck with a hammer, it scatters in an impressive way. Elemental mercury and mercury compounds also scatter through the environment. The major source of mercury emissions is coal-fired electric utilities. Mercury and mercury compounds occur extensively in nature and are found in coal. As coal burns, mercury compounds decompose to elemental mercury; then, as the flue gas cools and exits the plant, the majority of the mercury quickly oxidizes — probably catalytically, due to the presence of other metals in the gas — to its ionic form: Hg^{+2}. This form is water soluble, so it dissolves in rain and falls back to Earth.

If the rain is acidic, the solubility of the mercury increases. Since large quantities of rain fall into lakes and oceans, mercury levels build up. Bacteria in the water convert mercury to meth-ylmercury, its most toxic form. Methylmercury is fat soluble, so it accumulates in fish; the bigger the fish, the more of the substance it absorbs.

Smelters and incinerators also release mercury — crematoria included, because dental fillings are vaporized by the intense heat. Chlor-alkali plants, which produce chlorine and sodium hydroxide from salt, also release significant amounts of mercury. Land erosion is a contributing factor, because it permits mer-cury compounds in the soil to transfer into water. The flooding of previously wooded land is another factor. Vegetation and soil decompose and release mercury; furthermore, in such a bacteria-rich environment, mercury converts to methylmercury. Discarded thermometers and batteries release mercury as well.

Pregnant women, nursing mothers, and young children are particularly prone to mercury poisoning. Even low-level expo-sure can affect the developing brain and can have neurological and behavioral effects. And methylmercury can have an impact on the cardiovascular and immune systems. How much is too much? We don't yet have a consensus.

On the one hand, the FDA considers one part per million (ppm) of methylmercury in fish to be the limit. Shark, sword-fish, king mackerel, and tilefish can contain more than this. On the other hand, the Environmental Protection Association maintains that 0.25 ppm is the upper limit. If we adhere to this guideline, then white tuna (at 31 ppm) and light tuna (at 16 ppm) pose a problem. The species that canneries use for their light tuna contains less mercury than the albacore they use for white tuna because the fish are smaller. Current advice is that an adult should eat no more than one or two cans of tuna a week. A child should eat no more than one tuna sandwich a week, and before the age of five he or she should stick to flounder, haddock, sardines, crab, or shrimp.

125. What historic event in the annals of science took place on June 26, 2000?

At a press conference presided over by President Bill Clinton, J. Craig Venter of Celera Genomics (a private company) and Francis Collins of the National Institutes of Health's Human Genome Research Institute announced that the three billion or so chemical units that make up human DNA had been identified.

Locked into DNA are roughly fifty thousand specific sequences of these chemical units, which we refer to as genes. Some of these have already been identified in terms of their role, though the function of many others is unknown. But now that the whole genome — the collective noun that describes the complete set of a species' genes — has been sequenced, identifying specific genes will be much easier. And it will eventually mean that genetic analysis can be used to determine risk for specific diseases.

Genes also determine how a person will respond to drugs, so in the future doctors may keep files on their patients' DNA, allowing them to prescribe drugs that will work best with the fewest side effects. Drug development itself will become much more specific as we begin to comprehend the genetic causes of disease. Other possibilities are the prenatal correction of gene defects and the implantation of normal DNA sequences to correct genetic mutations.

Discovering the exact makeup of human DNA has been termed "cracking the genetic code," and that is what generated so much excitement in 2000.

126. There is often a black deposit on the inside surface of a burned-out lightbulb. What is this deposit?

———

It is metallic tungsten, the same stuff that the filament is made of. Here is what happens. Lightbulbs function according to the principle that passing an electric current through a resistance produces heat and light. The same principle applies to the toaster. From a lightbulb, we want the light; from a toaster, we want the heat.

The lightbulb's resistor is a thin filament of tungsten, which has the highest melting point of any metal. It also has a very low vapor pressure, and it does not readily evaporate — metals can indeed evaporate, mercury being a good example. However, hot tungsten does undergo a chemical reaction with oxygen, producing tungsten oxide. That's why a lightbulb has to be evacuated, or filled with an inert gas like argon or krypton. So, in a lightbulb the tungsten will not oxidize, but when the filament gets very hot some of the tungsten evaporates and condenses on the cooler glass surface, forming a black deposit. As more and more tungsten evaporates, the filament gets thinner, and the glass gets darker. The darkening of the glass is a signal that the filament will soon break and the bulb will blow.

127. What is conjugated linoleic acid, or CLA, and why are people so interested in it?

———

CLA is a component of certain fats, and it may have some interesting health properties. It seems incredible that we'd ever consider supplementing our diets with fat, but that's exactly what scientists may start encouraging us to do if further research bears out the early promise of conjugated linoleic acid.

This fat is mostly found in dairy products; whole milk and Cheddar cheese are the richest sources. Beef, lamb, and goat meat contain CLA as well. Bacteria in the guts of animals convert linoleic acid, a fatty acid found in animal feed, into CLA, which is stored in muscle and mammary tissue. We humans can't produce it ourselves, but we're very interested in it because research suggests that it is effective in fighting cancer, heart disease, diabetes, and weight gain. Rabbits fed a high-cholesterol diet are protected against heart disease if they also consume CLA. It lowers their triglycerides (blood fats) and reduces their LDL, the "bad" cholesterol. In rats, CLA works as an insulin-sensitizer. Type 2 diabetics can't produce enough insulin, and CLA may be a partial solution to their problem. In any case, CLA lowers triglycerides, which are always high in diabetics.

Perhaps the most captivating aspect of CLA is its role in controlling the body's muscle-to-fat ratio. In a three-month, placebo-controlled study, it significantly increased lean body mass in overweight patients. Over a twelve-week period, subjects taking 3.5 grams of CLA a day experienced a 1.7 kilo reduction in pure fat. According to one of the world's leading experts in this field, Dr. Michael Pariza of the University of Wisconsin, CLA's real potential has to do with its ability to prevent weight gain (as fat) after weight has been lost. Pariza himself takes three to four grams daily.

Not all CLA is created the same. Tonalin and Clarinol brands are manufactured to strict specifications. While there are several types of CLA, only the ones referred to as "cis-9, trans-11" and "cis-10, trans-12" isomers are biologically active.

The CLA connection to health is fascinating, but that's no reason to start gorging on meat and high-fat dairy products. The benefits that study subjects achieved are unlikely to be achieved to the same degree through diet. The diet of the average North American includes about one hundred milligrams of CLA — much less than the dose Dr. Pariza is taking. So, if further research does prove CLA's benefits, then supplements will be the way to go. Or perhaps scientists will add sunflower oil or safflower oil to cow feed. Or maybe they'll genetically engineer cows to produce more CLA.

128. Why did the Maitland, Florida, city council ban the planting of angel's trumpet, an ornamental plant?

Angel's trumpet contains atropine, a chemical that blocks the activity of acetylcholine, one of the body's important neurotransmitters. Atropine dries up bodily fluids — saliva, tears, mucus, phlegm, and urine — and it induces hallucinations. In a sufficiently high dose, it can kill.

Teenagers have tried to exploit the plant's hallucinogenic properties by drinking tea brewed from its leaves. The resulting rash of poisonings triggered the Maitland ban. There is no doubt that atropine can kill. Indeed, as a poison, it is almost perfect, because it breaks down in the body very quickly and is difficult to detect.

An Edinburgh biochemistry professor named Paul Agutter was well aware of this. He tried to murder his wife by slipping some atropine into the tonic water she mixed with her gin. The bitter flavor of the tonic disguised the taste of the atropine. To throw police off his trail, Agutter adulterated some bottles of tonic water in his local supermarket with small doses of atropine, creating the impression that a mad poisoner was on the loose. But Agutter was unlucky. An anesthetist's wife purchased one of the tainted bottles. When she and her son became ill, her husband recognized the symptoms as atropine poisoning and informed the hospital.

Within the next few days, five other tainted-tonic drinkers ended up in the same hospital. All survived after being treated for atropine poisoning. The police tested their tonic water and discovered atropine, but they found that the bottle belonging to Mrs. Agutter contained far more of the poison. Agutter finally confessed and was sentenced to twelve years in prison. Why did he want to murder his wife? The age-old reason. He wanted to inherit her family fortune and then marry his mistress.

129. The Italians have a saying that one should eat cherries only before St. John the Baptist's Day. Why?

The saying is based on the notion that after June 24 cherries harbor worms. There's actually some truth to this. The worms in question are the larvae of the cherry fruit fly, which hatch from eggs laid by adult flies.

Around the end of May, depending on weather conditions, pupae, buried in the soil beneath the cherry trees the year before, develop into fruit flies. They emerge from the ground and fly up into the trees, where they feed on bird droppings and tree resins. In about five days, they lay their eggs under the skin of the unripe, straw-colored cherries. In about ten days, the eggs hatch; within ten more days, they are full-sized larvae. This brings us to about the third week of June.

So, before this time, you're unlikely to bite into a worm as you enjoy fresh cherries. Later on, however, you may not be so lucky, because the adult flies keep emerging from the ground for many weeks, and the cycle continues. What happens to the larvae? After they have gorged themselves on cherry flesh, they drop to the ground and burrow in. They metamorphose into pupae and lie dormant in the ground for about ten months, only to reemerge as adults. Of course, commercially produced fruit is harvested from trees that have been sprayed, allowing us to eat worm-free cherries way past June 24.

130. Scientists from Simon Fraser University in British Columbia cast six pig carcasses into Lake Ontario. Why?

While we don't like to dwell on it, the body of a pig is very similar to that of a human. The researchers were interested in collecting data on how human bodies decompose in water, because such data is valuable in determining time of death. Divers checked the carcasses several times a week, removed samples for analysis, and videotaped the remains. Not a pretty picture.

The pig carcasses, which weighed from 50 to 150 pounds, were donated by a local butcher. So, what kind of researcher gets involved in such work? A forensic entomologist — a scientist who uses his or her knowledge of insects to solve legal matters.

Here's an example. One Kathleen McClung was murdered during the night of June 20–21, 1969. The coroner performed a postmortem examination on June 24 and collected two tubes of maggots. The first set came from the victim's mouth. They were small and dead and difficult to identify. The second set came from the polyethylene bag that was wrapped around the victim's head when she arrived at the mortuary. These maggots were alive. Investigators fed them raw beef and obtained eight pupae between July 4 and 8. All eight hatched as flies between July 18 and 23, and they proved to be the common *Calliphora vicina*. By looking at the pupation dates, investigators determined that the eggs had been laid between June 21 and 24. From this, they were able to confirm the date of McClung's death.

131. The Chinese are studying an extremely toxic substance isolated from a fish to see whether it can ease cancer-related pain and the symptoms of heroin withdrawal. What kind of fish?

The puffer fish. This spiny fish, which inflates like a balloon when threatened, harbors tetrodotoxin, one of the most powerful poisons ever discovered.

The toxin occurs only in certain parts of the fish, and the fish becomes edible when these parts are removed. Indeed, the Japanese consider a dish made from the puffer fish, or fugu, a great delicacy. Chefs who are licensed to prepare it must first undergo extensive training in puffer fish evisceration. Still, there are a couple of fugu deaths in Japan each year. The toxin works by interfering with nerve conduction and paralyzing critical systems in the body. Nerve conduction depends on an influx of sodium ions into nerves from surrounding fluid via sodium channels. Tetrodotoxin blocks these channels.

Since the early 1990s, researchers at Beijing Medical University have been investigating the possibility that, in the proper dose, tetrodotoxin interferes sufficiently with nerve conduction to block pain while still preserving life. They have found the substance to be effective against cancer pain. A group of Chinese cancer patients who were on pethidine hydrochloride (a morphine analogue) continuously and experiencing little relief all claimed to be pain-free after three days of twice-daily tetrodotoxin injections. Tetrodotoxin has also been used against heroin addiction successfully — heroin addicts have kicked the habit after just three to seven days of injections.

This is another excellent example of how chemicals are neither good nor bad — their effects depend entirely on the dose and the context. Dose is particularly important when it comes to tetrodotoxin — badly prepared fugu isn't a reliable way to treat pain.

132. At one time, in France, some wine producers had a sideline that involved placing fermented grape skins between thin copper sheets. What was it?

––––

The making of verdigris. Verdigris, a greenish-blue substance, is an effective fungicide and a paint ingredient. Chemically speaking, it's a complex of copper acetate and copper hydroxide. The name derives from "verte de Grèce," or "green of Greece," because the ancient Greeks were probably the first to make it by hanging copper plates over hot vinegar in a sealed pot until a green crust formed on the copper.

Fermented grape skins contain alcohol along with acetic acid, which is the product of the reaction of alcohol with oxygen of the air. The acetic acid reacts with copper to produce copper acetate. This substance is effective against fungi that attack grapes. Copper acetate can also be manufactured by reacting copper sulfate with calcium acetate or barium acetate. Both calcium sulfate and barium sulfate are insoluble and precipitate out; copper acetate stays in solution.

Producing verdigris artificially on a copper object isn't an easy task. Artists usually resort to using acrylic paints to mimic the natural reaction of copper with acidic compounds in the air. By the way, the patina on copper roofs is a mixture of copper carbonate and copper sulfate.

133. What plant food, a staple in sub-Saharan Africa, can cause death and paralysis if it's not properly processed?

––––

Cassava, also known as tapioca. The plant is a tuber, much like a potato, and was introduced to Africa from South America by the Portuguese. Like many plants, some varieties of cassava harbor a chemical that protects it from predators. It's called linamarin, and it works by releasing cyanide. When an animal attacker's bite damages the tissues of a cassava, the plant's storage cells release an enzyme that reacts with linamarin, liberating cyanide. Cyanide's bitter taste is usually enough to make the animal retreat, but, if the animal eats enough cassava, it will die. So will a human.

Over the years, through trial and error, cassava eaters have identified a variety of processes that render the tuber safe. South American native peoples who spat bitter-tasting cassava back into their food bowls and then sampled it again later discovered that the bitterness had vanished. That's because salivary enzymes can also cause linamarin to break down, liberating hydrogen cyanide, which then escapes into the air. In Africa, the traditional method of detoxifying cassava is to pile cassava roots and allow them to ferment in the sun for about seven days. This also releases the enzyme needed to degrade linamarin.

Still, cases of cassava poisoning occur relatively often. In times of drought, when people are very hungry, some do not process their cassava long enough. This can cause "konzo," an irreversible paralysis of the legs due to chronic, sublethal cassava ingestion. A test strip is now available to test for cyanide in cassava. The strip is treated with a solution of picric acid, which turns brown in the presence of cyanide.

134. You simmer a pad of soapless steel wool in vinegar for a few minutes. Then you remove the pad and add a little hydrogen peroxide. Finally, you pour this solution into a cup of tea. What happens next, and what have you made?

The tea turns black, and you've made a simple ink. Tea contains tannic acid, a substance that forms when tea leaves are fermented and dried. Various polyphenols present in tea leaves combine to form the acid, which can react with iron to form a dark complex suitable for use as ink. But tannic acid doesn't react with iron in its elemental state. The iron has to be in the form of ferric ions, meaning that each iron atom has to give up three electrons and assume a positive charge.

When iron is boiled in acetic acid, or vinegar, it converts to ferrous acetate. Some of the hydrogen atoms in the vinegar steal electrons from the iron to form hydrogen gas. This leaves iron with a plus-two charge. Hydrogen peroxide takes another electron away, leaving behind ferric ions; in the process, the hydrogen peroxide converts to water. The ferric ions react with tannic acid to produce ink.

Ferric tannate is the major ingredient of black ink, and acid destroys it. That's how old-fashioned ink eradicators worked — you sprinkled a little acid on your error and hoped it would destroy the ink before it destroyed the paper. Careful blotting was clearly necessary.

135. Why does adding lemon juice to tea lighten its color?

The chemistry of ink eradicators, discussed in the answer to the previous question, comes into play here as well. Acids — such as citric acid in lemon juice — break down the iron-tannate complex. But since we don't add steel wool to our tea, where does the iron come from? It comes from the water.

Now, we're not talking chunks of floating metal here — we're talking iron compounds dissolved in water. Soil contains a variety of iron compounds, which dissolve as water percolates through it. The color of tea varies according to the iron content of the water it's brewed with. Some water filter manufacturers advertise that their products will give you lighter-colored tea. While this is true, it's of no real consequence. All it means is that the filters remove iron from the water, preventing the iron-tannate complex from forming.

While we're discussing tea, let's look at an issue involving its health benefits. Catechins — tea's beneficial compounds — can complex with casein, a milk protein. That's why some people are concerned about the effects of adding milk to their tea. But Dutch researchers looked at the blood catechin concentrations of a group of subjects after they drank green tea, black tea, and black tea with milk. Twelve subjects consumed three grams (roughly six cups) of tea a day. They reached peak catechin levels after 2.5 hours, and the addition of milk made no difference. Actually, the half-life for blood clearance was 4.8 hours for green tea, 6.9 for black, and 8.6 for black tea with milk. So, if anything, the milk seems to help.

Tea may have yet another health benefit. Japanese researchers examined the use of teabags to cure "sick-building syndrome." They scattered them around buildings to absorb gases that were making people sick. The concentration of formaldehyde, which can outgas from particleboard, was decreased by over 60 percent. It seems that tea may be healthy in more ways than one.

136. In 1948, Swiss engineer George de Mestral went hunting in the Alps. He passed through a burdock thicket, and this prompted one of the most useful inventions of the twentieth century. What was it?

———

Velcro. The burdock plant propagates by releasing burdock seed heads, better known as cockleburs. It's nature's way of ensuring that the seeds spread. Any animal that rubs against the plant becomes a potential vehicle for seed dissemination.

During his hunting expedition, de Mestral became annoyed by the cockleburs that stuck to his clothes and to his dog. As he tried to pull them off, he wondered what made them so tenacious. He slid a sample under his microscope and noticed many little hooks on the surface of the cocklebur. These hooks had become entangled in the loops of thread that made up the fabric of his clothes and in the coiled hairs of his dog's coat.

De Mestral understood that a reusable fastener would have enormous commercial appeal, and he became preoccupied with finding a way to replicate the cocklebur phenomenon. It took him eight years to come up with Velcro, which he named after the French for velvet (*velours*) and hook (*crochet*). Eventually, the engineer discovered that nylon was the best material for his venture. It didn't break down, it didn't rot, it didn't attract mold, and it could be produced in threads of various thicknesses. De Mestral had trouble creating hooks until he realized that he could do it by cutting loops in half. He later strengthened the filament by blending polyester with the nylon.

Velcro now has an amazing array of uses. On each space shuttle, there are ten thousand inches of Velcro tape; it's made of Teflon loops, polyester hooks, and glass backing. Inside the helmet of each astronaut is a small Velcro patch — it serves as a nose scratcher during space walks. Manufacturers have

even developed silent Velcro in response to a request from the U.S. Army for a fastener without the rip sound, which could betray a soldier's position to the enemy. They were able to reduce the noise by over 95 percent, but how they did it is a military secret.

Velcro-covered suits are now available for those who engage in the curious activity of Velcro jumping. Participants don the suits, take a running jump, and hurl themselves as high as possible at a Velcro-covered wall.

But not every Velcro application has worked. The "Worm Collar" was not a success. It was a cylindrical strip that fishing enthusiasts were meant to slip onto the bodies of their live bait; the idea was that it would snag the teeth of striking bass. A strap-on device for impotent men also flopped.

137. During a celebrated 1893 murder trial, prosecutors conducted an experiment before the jury. They administered a lethal dose of morphine to a cat, and, as its life ebbed away, they applied a few drops of belladonna extract to its eyes. Why?

———

The prosecutors wanted to prove that one can reverse the pinpointing effect that morphine has on the pupils of the eye by using atropine, which is found in the belladonna plant.

Dr. Robert Buchanan was a physician whose wife suddenly sickened and died. The medical examiner concluded that the cause of death was a brain hemorrhage. This didn't sit well with the former lover of the woman Buchanan was having an affair with. The doctor had done away with his wife, the man insisted, probably with morphine. "Couldn't be," said the medical examiner, "the victim didn't have pinpoint pupils."

When the *New York World* started to question the handling of the case, the coroner gave in to the pressure and ordered an exhumation. The lady's remains were found to contain morphine in a dose sufficient to cause death, and Dr. Buchanan went on trial for murder.

His undoing was a conversation reported by a witness. The doctor, it seems, had railed against another accused morphine poisoner, calling him an incompetent. That poisoner had been undone by the fact that his victim had pinpoint pupils, and Buchanan had declared the fellow a fool for neglecting to apply belladonna. When this story surfaced, another witness recalled Dr. Buchanan putting drops in his wife's eyes before she died. The good doctor was executed in the same electric chair in which the "bungler" had met his end two years earlier.

138. The rulers of ancient Greece, Persia, and Rome often traveled to Delphi in Greece to seek the oracle's counsel on matters of war and love. What is the connection between the rantings of the pythia, the priestess who spoke on behalf of the god Apollo, and tomatoes?

_____

The Delphi oracle was active from roughly 1600 B.C. to 392 A.D., when a Christian king put the enterprise out of business. The oracle's shrine on the slopes of Mount Parnassos was an imposing structure supported by numerous columns. But the essence of the temple was a small chamber, called the *adytum*, hidden below the floor.

The *pythia* went to this chamber to commune with Apollo. According to ancient Greek writers, she sat on a tripod and inhaled the *pneuma*, or fumes, that emanated from fissures in the ground. This put her into a trance, and she would babble answers to the questions posed to her. Modern research indicates that there's something to the delirium story. On Parnassos, there is plenty of bitumen in the ground, and, when it's heated by the friction of moving rocks, it releases a variety of hydrocarbons, including ethylene. This gas can cause lightheadedness and euphoria; at one time, it was even used as a surgical anesthetic. And it can certainly trigger violent frenzy and delirium. So, perhaps the *pythia* was high on ethylene when she delivered to Oedipus this self-fulfilling prophecy: "Away from the shrine, wretch! You will kill your father and marry your mother!"

And the tomato connection? Ethylene is a plant hormone that causes tomatoes to ripen. Tomatoes are often picked green and ripened with ethylene gas. But I wouldn't count on predicting the future with ketchup.

139. According to a Greek legend, a sheepdog belonging to Hercules bit into a mollusk while walking along a beach in the ancient city of Tyre. What important industry did this launch?

The dye industry. This tale refers to one of the most important dyes of the ancient world: Tyrian purple. Species of mollusk known as *Murex brandaris* and *Murex trunculus* produce a purple color, which supposedly stained Hercules's dog.

Hercules's dog was mythical but Tyrian purple certainly was real. Historians tell us that by 1500 B.C. Tyre was a prosperous dye-producing center. The extraction process involved crushing the shellfish, salting them, and boiling them for ten days. Huge quantities of mollusks were needed to produce any significant amount of dye, so it was a very expensive product. Only the rich could afford it, and Tyrian purple came to symbolize wealth, high achievement, and respect. Cleopatra adorned her barge with purple fabrics, and Julius Caesar decreed that only the emperor and members of his entourage could wear Tyrian purple. All of this gave rise to the expression "born to the purple," which means born into a noble household. The fringes of the tallit, the traditional Jewish prayer shawl, were also dyed with the same mollusk extract.

Tyre gained worldwide renown for the beautiful dye it produced and for something else as well. It became famous as the world's foulest-smelling city due to the huge piles of rotting shellfish.

140. How are Beethoven and plumbing related?

Our word *plumbing* derives from the Latin *plumbum*, meaning "lead." This is why the element's chemical symbol is Pb. The

Romans used lead to make water pipes, hence the derivation of the word. Lead is a toxic metal, and those who ingest it in even small amounts experience a wide range of symptoms. These include abdominal pain, nausea, slowed reflexes, weakness, vertigo, tremors, loss of appetite, depression, confusion, irritability, anxiety, and — in children — learning difficulties. And guess what symptoms Beethoven suffered from?

From the age of twenty, the composer complained constantly of stomachaches, digestive problems, depression, and irritability. His doctors were at a loss for an explanation. Now, however, we finally may have one. When Beethoven died in 1827, at the age of fifty-nine, a young music student clipped a lock of his hair as a souvenir. It was passed down through the student's family and sold at auction in 1994.

The buyers subjected the hair to chemical analysis to see what they could discover about the great man. Did he die of syphilis, as many have claimed? If so, he would have been treated with mercurial drugs, remnants of which would show up in the hair. The testers found no mercury. Neither did they find opiate residues, suggesting, perhaps, that Beethoven shunned painkillers in an effort to keep his mind clear for writing music. But what surprised the testers was the high level of lead in the composer's hair. About sixty parts per million — which is roughly a hundred times the amount an average person would have today. We can only guess how Beethoven slowly poisoned himself. Perhaps he drank regularly from lead mugs, which were common at the time. Maybe he drank water that flowed through lead pipes. Or maybe he was somehow exposed to lead-based paints. Whatever the solution to this mystery may be, there's a good chance that one of the world's greatest composers was silenced by plumbing.

And here's another Beethoven mystery that will never be solved. How did the man compose his brilliant pieces, given the

fact that for most of his life he heard no sounds. Ludwig van
Beethoven was deaf!

141. What is the link between Florida shellfish contaminated by a natural toxin and the Sahara Desert?

———

The link is algae, specifically *Karenia brevis*, which is responsible for a phenomenon called red tide. Outbreaks of red tide
occur periodically in the ocean, often off the coast of Florida.
They happen when algae multiply rapidly, coloring the surface
of the water red.

Red tide has struck throughout history. One outbreak that
occurred in ancient Egypt may even explain the biblical story
of Moses turning the Nile River red. Mythology related to
red tide crops up everywhere. During a serious outbreak that
occurred as World War I raged, sailors said that the red water
was reflecting the flames of hell — the gates of the underworld
were open to receive multitudes of new tenants.

Today, scientists don't subscribe to this theory. They know
that the *Karenia brevis* blooms when there's a source of nitrogen
fertilizer in the water, such as urea or ammonium ions. Under
normal conditions, these aren't present, but they are formed with
the help of another type of algae, called *Trichodesmium*. It has
the ability to "fix" nitrogen — meaning it can convert nitrogen
in the air into soluble and usable nitrogen compounds.

This species of algae is readily fertilized by iron in the water
— iron that may have blown in from the Sahara! Researchers
have shown that ferocious storm systems scoop up dust and
carry it across the Atlantic. They have linked the occurrence
of these dust storms with a hundred-fold growth of *Trichodes-*

mium and the onset of red tide. Unfortunately, the red tide algae produce a toxin that is concentrated in shellfish and can cause paralysis and even death.

142. The world's oceans are filled with single-cell organisms called algae. If they suddenly vanished, so would we. Why?

There is no life without oxygen, and most of the oxygen we breathe is produced by algae through photosynthesis. Algae take in water and carbon dioxide and convert them to glucose and oxygen.

Of course, all plants do this. That's why people think we need forests to supply us with oxygen. This is not so. While trees do photosynthesize and generate oxygen, they also consume oxygen. Photosynthesis takes place during the day, but during the dark hours trees and plants undergo the process of respiration — they take in oxygen. Just like humans, they need oxygen to fuel their continuous metabolic reactions. Furthermore, when trees and plants die, they biodegrade, and this requires oxygen as well. What it all comes down to is that a plant, over its life cycle, consumes almost as much oxygen as it produces. There are plenty of reasons not to devastate our forests, but oxygen preservation isn't one of them.

Algae also use up oxygen, but there are far more algae in the world's waters than there are plants on good old terra firma. This means that the small excess of oxygen algae produce when you take away what they consume translates into a massive amount of oxygen. And that is where the 20 percent of oxygen in our air comes from.

143. The year 2001 was the one hundredth anniversary of what important event in the history of science?

———

The awarding of the first Nobel Prize. The Nobel Prize is the most prestigious award in science, and was first awarded in 1901, in accordance with the will of Alfred Nobel, the Swedish inventor who died in 1896.

Nobel's greatest scientific contribution was inventing dynamite, which was basically a type of clay, called kieselguhr, saturated with blasting oil, better known as nitroglycerin. Ascanio Sobrero of the University of Turin had made nitroglycerin in 1845 by treating glycerol with a mixture of nitric and sulfuric acids. Nitroglycerin's blasting power was tremendous, but the substance was difficult to make and to work with.

The synthesis of nitroglycerin was a dangerous undertaking. The reaction had to be watched carefully, because it produced a lot of heat and could easily get out of control. At a nitro plant in Ardeer, Scotland, workers assigned to watch the stuff were forced to sit on one-legged stools so they wouldn't doze off! Nitroglycerin was also risky to transport and hard to detonate. Nobel invented a detonating cap made of mercury fulminate, which proved extremely useful. But what brought him fame and fortune was his idea of mixing nitro with kieselguhr to make dynamite. Dynamite was safer, because you could only explode it by igniting it with a fuse.

Nobel was concerned that some might try to use his discovery to the detriment of humankind. He worked to foster the peaceful implementation of science by establishing annual awards in the fields of chemistry, physics, and medicine. He also believed that there was more to life than science, so he established awards for literature and peace as well. And, in 1968, the Nobel Foundation added a prize for economics.

Nitro proved to be valuable not only in making explosives but also in treating angina. In very small doses, nitroglycerin can relieve the crushing chest pain associated with partially blocked coronary arteries. Nobel himself suffered from this condition in his later years, and his doctors prescribed nitroglycerin. Indeed, without the drug, he wouldn't have been able to sit at his desk and make out his will, in which he laid out the provisions for the Nobel Prizes. "It sounds like an irony of fate that I should be ordered to take nitroglycerin internally," he remarked. Indeed it was.

144. In 1879, a worker at Procter and Gamble went for a lunch break and forgot to turn off the stirrer of the large pot in which he was blending chemicals. When he returned, he was surprised by what he saw. What popular new product had he accidentally discovered?

Floating soap. Soap's essential ingredients are fat and an alkaline material. Wood ashes were the original alkali source, but producing large amounts was a problem. Then, in 1789, along came Nicolas Leblanc, the French chemist who introduced a process whereby soda, the necessary ingredient from the wood ashes, was mass-produced from brine. Now soap manufacturers had plenty of soda to mix with the fat in their big kettles.

In America, Procter and Gamble got into soap manufacturing in a big way when cottonseed oil became available as a cheap source of fat. The forgetful worker who left his stirring machine running had been mixing the oil with soda. The soap that formed in his absence was lighter than usual, because the machine had beaten plenty of air into it — in fact, it was so light that it floated. Procter and Gamble was impressed and decided to try selling it. Soon, letters from intrigued consumers were pouring in — everyone wanted more of the miraculous product that made it unnecessary to grope the bottom of the bathtub for that elusive bar of soap.

Eager to capitalize on the demand, the company quickly got down to the business of promoting Ivory, the floating soap. Chemists analyzing the soap to ensure that the manufacturer could reproduce it determined that the long stirring time had caused almost all of the fat and soda to be converted to soap, leaving behind little in the way of unused reagents. Ivory, in fact, turned out to be "99 and 44/100 percent pure." Another advertising slogan was born. But this figure does not take into account the ingredient that makes Ivory float — which, of course, is simply air.

145. What is the connection between white paper and skywriting?

Titanium dioxide. Paper is made from wood pulp, which isn't white. Manufacturers employ a variety of bleaching agents to lighten the color of their products, but to make high-quality white paper they must add titanium dioxide to the mix. This opaque white pigment is used to make white paint, whitewall tires, and sunblocks. It is the main ingredient in liquid paper, which, of course, is great for covering up mistakes. It can also be used for skywriting. And therein lies some interesting chemistry.

Titanium dioxide is made from a precursor, titanium tetra-chloride. This liquid reacts readily with moisture to produce titanium dioxide and hydrochloric acid. As skywriters release the liquid from containers attached to their airplanes, it reacts with moisture in the air to form finely powdered titanium dioxide. Since the tiny particles take a little while to disperse, skywriters have time to get their messages across. Titanium dioxide is nontoxic — it's even used as a food colorant — but the stuff is difficult to manufacture. The process used to make it industrially is the same one the skywriters use. It generates a great deal of hydrochloric acid, and manufacturers have to deal with that.

A novel function for titanium dioxide is attracting a lot of attention. The substance may be just the right stuff for cleaning tiles, eliminating bacteria, reducing smells, and even cleaning water. How can it do all that? Through some neat photochemistry. When light hits titanium dioxide, it dislodges electrons. If oxygen and water are around, these electrons combine with them to produce hydroxyl and superoxide radicals — very active chemical species that destroy organic molecules. That's how titanium dioxide eliminates stains and reduces odors.

The Japanese have already conducted some experiments involving urinals. Researchers coated the ceramic of several urinals with a layer of titanium dioxide no thicker than a human hair and left others untouched. After a month, the untreated urinals had yellow blotches, but the treated ones were still white. Titanium dioxide may also be a boon for hospitals. It can be incorporated into wall and floor tiles to kill bacteria on contact. And even fastidious housekeepers may benefit. It seems that a thin layer of titanium dioxide keeps dust from sticking to chandeliers. The compound may even prove to be effective in water filters.

All of this is very interesting, but most people know nothing about it. Maybe some clever chap will start promoting the benefits of titanium dioxide with a little skywriting.

146. What's the difference between regular candles and dripless ones?

———

Candles, one of our oldest tools for generating light, were the result of an extended search for materials that burn for a long time. Nobody knows who first determined that beeswax, beef tallow, and spermaceti from whales all burn well, but it must have occurred to many. Countless individuals must have observed that the fatty parts of animals would sometimes catch fire as they cooked.

Until the early part of the nineteenth century, people made candles by coating wicks with beeswax, tallow, or spermaceti. Then, as the chemical industry got going, stearic acid became more readily available. It is derived from sodium stearate, better known as soap. Treatment of animal fats with sodium hydroxide results in sodium stearate, which can be treated with an acid to yield stearic acid. This substance, people discovered, burned well and could be produced cheaply. It was great for candles, except for one thing: it was brittle. Someone solved this problem by adding paraffin, a mixture of long hydrocarbon molecules isolated from petroleum.

Now a new difficulty arose. Paraffin made the candle wax melt more easily — it dripped down the sides of the candle, forming "waxicles." To make a dripless candle, we need a wax that has a melting point high enough to prevent the candle flame from melting the edges. The solution is to increase the amount of stearic acid. Of course, we can just burn candles with edges that are far from the flame — a big candle won't drip.

Here's a simple dripless-candle-making trick you can try. Pour some water into a bowl and add a couple of tablespoons of salt. Soak ordinary candles in the water for a couple of hours, and presto! You have dripless candles. The wax absorbs the salt, which raises its melting point.

But if you want a really interesting candle, you'll have to get hold of a dried stormy petrel, a Shetland Islands bird that has a very high fat content. Islanders fashion candles out of them — they thread a wick through a dead bird's beak, fix its feet in clay, and then set the thing alight. But I bet the bird drips.

147. How do antifog sprays for eyeglasses work?

The fog that forms on eyeglasses consists of small droplets of water. Water vapor from the air condenses on the cold surface of the glass, and the droplets are thick enough to distort one's vision. There are two ways to approach this fog problem. First, we can heat the glass — some expensive hotels have heated, "fog proof" bathroom mirrors. Second, we can prevent the condensed water from forming droplets.

This is achieved by lowering the water's surface tension, the force with which water molecules are attracted to each other. Water beads in the first place because the attraction water molecules have for each other is greater than their attraction to the surface. The idea, then, is to introduce a substance that will get between the water molecules and prevent them from sticking to each other. Instead of forming droplets, the water will spread into a thin film, which won't obstruct vision. Several substances can perform the job. Isopropanol, or rubbing alcohol, is the most common, but a variety of detergents can also help us see things more clearly.

148. How come skiers say that artificial snow feels different from regular snow?

––––

Because artificial snow is different. It isn't really snow. Snowmaking machines actually manufacture tiny beads of ice, each about one-ten-thousandth of an inch in diameter. The machines work by spraying water and compressed air through a hose. When a gas, such as compressed air, rapidly expands, it cools. This cooling helps to freeze the water. As the water freezes, heat is released. This sounds paradoxical until we realize that it takes heat to melt ice, heat that therefore must be liberated when water freezes. Incidentally, this is why snowmaking pipes are always positioned high in the air. If the machine made its snow too close to the ground, then the heat released through the freezing process would actually warm up the ground and melt some of the snow.

Farmers who spray their crops with water when a freeze is forecast are abiding by the same principle. Many people assume that the thin layer of ice that the spray produces acts as insulation — not so. In fact, the freezing water liberates enough heat to keep the water inside the crops from freezing.

So, why does skiing on artificial snow feel different? When we ski, the pressure of our skis on the snow's surface causes it to melt, increasing its slipperiness. But real snowflakes and artificial ice beads have different contact surfaces, and their temperatures can be different, so the rate at which they melt under pressure varies. With my meager ski talent, I can't say that I've noticed much of a difference other than the fact that it hurts more when I fall on the artificial stuff.

149. How does baking soda absorb odors in the fridge?

————

Smells occur when volatile compounds stimulate receptors in our noses. There is a tremendous variety of such compounds, and they display a great diversity of molecular structures. But many of the smells we encounter in our fridges are due to volatile fatty acids — for example, when butter goes rancid, it releases butyric acid, which has a particularly foul smell. Bases neutralize acids. Baking soda, or sodium bicarbonate, is a base. It reacts with butyric acid to form sodium butyrate, which has no smell because it isn't volatile.

Most fridge smells don't originate directly in food. Rather, a complex ecology of bacteria and household mildew like *Aspergillus* produces the nasty volatile compounds. These mildews and bacteria like to dine on the various nutrients available in a fridge, and they create a range of acids with disturbing smells. *Pseudomonas* bacteria emit a mildew-like smell, and they congregate in freezers and on dishcloths and towels. In the fridge, they coexist with microbes that inhibit the bacteria or break down the smell, but these microbes cannot live in the freezer. So, in the freezer, the bacteria have less competition — they multiply and produce a smell.

Baking soda can help, but only if you give it some space to work. If you just open a corner of the box, nothing will happen. The baking soda needs to cover a large surface area, so spreading some on a plate is the best way to go. If this doesn't solve your smell problem, then try washing the inside of your fridge with diluted bleach, followed by hydrogen peroxide to get rid of the chlorine smell. Rinse with a solution of bicarbonate, which changes residual smelly volatile fatty acids into sodium salts. If the situation is desperate, dust the interior of the freezer with a little flowers of sulfur or zinc oxide, both of which will discourage microbial growth. They're not poisonous, but you should still wear gloves when handling them.

What can you do with the used baking soda? I wouldn't suggest using it in your baking, because heat will liberate the smells. Give it to the kids to build a volcano. Tell them to combine the soda with vinegar and watch it fizz. This may release those bad smells, but you'd expect a volcano to be smelly.

150. Where does the notion that green potatoes are dangerous to eat come from?

Potatoes contain compounds that belong to the alkaloid family. Two of these, solanine and chaconine, are toxic in a high enough concentration. They can create digestive distress and interfere with nerve transmission.

Exposure to light or storage conditions that stress the potato stimulate the formation of these compounds. Temperatures that are either too cold or too hot stress potatoes. So do bugs. A University of Guelph study showed that potato plants stressed with Colorado potato beetles and leafhoppers had significantly higher glycoalkaloid concentrations. Judicious use of pesticides may therefore increase safety.

Solanine and chaconine are probably the most widely consumed natural toxins in North America. These alkaloids are colorless, but their production parallels the production of chlorophyll — which, of course, is green. That's why green potatoes are suspect. Sometimes, when their concentration is high enough, solanine and chaconine will make your tongue burn. Potato sprouts are rife with alkaloids, so you should always cut them off.

The alkaloids are not destroyed by heat, but, since they are mostly concentrated very near the potato's surface, peeling pretty much eliminates the problem. However, if you like to eat potato skins, don't worry about it too much — you would have to eat over a pound before symptoms would appear. So, solanine and chaconine don't present a great risk, but don't go seeking out green potatoes to eat.

151. Why did steakhouses stop serving steaks on wooden plates?

When I was growing up in Montreal, I looked forward to my family's annual pilgrimage to a great local steakhouse where we would be served gigantic slabs of meat hanging off wooden plates. Then, all of a sudden, the health authorities decided that these plates were not hygienic because the meat juices seeped into the wood, and this fostered the growth of bacteria. The wooden plates were banned, and after that steaks somehow never held the same appeal.

But science marches on. Recently, researchers got around to testing the theory that wooden plates and cutting boards harbor bacteria. Much to their surprise, when they inoculated the wood with bacterial cultures, they found that the wood had a decidedly antibacterial effect. Apparently, as yet unidentified compounds that occur naturally in wood do away with microbes. So, now we know that wooden cutting boards are not more dangerous than plastic ones. In fact, studies have shown that plastic develops grooves in which bacteria grow, and even vigorous washing won't dislodge them.

Although we've let wooden boards out of the doghouse, it's still a good idea to keep them as clean as possible. Rubbing the board with salt works — it kills bacteria by dehydrating them. Scrubbing well with soap also works. But the best way to disinfect a board is to put it in the microwave oven for a minute or two. Keep an eye on it to make sure that it doesn't char. Now, if steakhouses would only bring back those wooden plates...

152. How do you make a marshmallow?

Believe it or not, marshmallows were inspired by a plant. The marshmallow is a perennial that grows to a height of about four feet, and its root contains a great deal of polysaccharides we refer to as "mucilage." When added to sugary foods, mucilage inhibits crystallization. The root actually resembles lung tissue, inspiring people in days gone by to use it as a treatment for lung conditions. They would prepare an extract from the root, which they sweetened and aerated. Today, some people still use marshmallow tea to soothe inflammation.

Marshmallow candy once contained real marshmallow root, but not anymore. Nowadays, candy manufacturers concoct the soft treats with a blend of sugar syrup and gelatin. Of course, all candy is based on sugars, and all candy makers must understand the relation between the boiling point of a sugar solution and its sugar concentration. Solutions always have a higher boiling point than pure liquids, so a sugar solution always boils above 100°C (212°F). The more sugar it contains, the higher the boiling point. As the solution heats, water evaporates, and the solution becomes more and more concentrated.

The sugar concentration determines the structure of the candy after the syrup cools. For example, an 85-percent sugar solution forms when the mixture boils to about 105°C (221°F), and when it cools it will be fudge. To make marshmallows, candy makers must boil the solution to about 110°C (230°F). Then they add corn syrup, because its long chains of glucose get in the way of the sugar molecules looking to add on to sugar crystals. By inhibiting crystallization, the syrup keeps the candy from hardening — after all, no one wants a hard marshmallow.

The key to marshmallows, however, is gelatin, a protein that coagulates when beaten, just like the protein in egg whites. In fact, some marshmallows are made with egg whites. The coagulation occurs at the air-liquid interface as the sugar-protein mixture is whipped, and it stabilizes the air bubbles that form in the mix. When you plop a marshmallow into your hot chocolate, it doesn't melt — it dissolves. Since sugar is water soluble, the confection begins to disintegrate. It will not dissolve completely, because the coagulated gelatin is not water soluble.

If you want your marshmallow to melt, you have to impale it on a stick and hold it over a flame. The sugar will melt — and then some. It will actually begin to break apart chemically, emitting the delicious fragrance of caramel. But don't eat too many burned marshmallows, since the carbonized part contains molecules that are carcinogenic in animals. Still, how often do we eat marshmallows around a campfire, anyway? And when it comes to enjoying marshmallows in hot chocolate, our only worry is cavities.

153. Why is milk white and not green, like grass?

Questions about milk are always interesting, because milk is our first food. It is the food that sustains us completely in early life, and it must therefore contain all the required nutrients: water, proteins, fats, sugar, vitamins, and minerals. Chlorophyll, the green coloring agent in plants, is not a required human nutrient, so the evolutionary process has not resulted in chlorophyll being incorporated into mother's milk.

This is not to say that milk has no colored substances. It does. Milk contains riboflavin, or vitamin B2, an important enzyme cofactor that has a greenish-yellow hue. Riboflavin is water soluble, so its color is easiest to see in the whey that forms when the liquid portion of milk is separated from the solids, as it is in cheese making. The fat globules in milk have a yellow tinge due to carotenoids, such as beta-carotene, which come from the diet of the milk source. Beta-carotene has nutritional importance because it is a precursor to vitamin A. Summer cow's milk is yellower that winter cow's milk because of what the cows have been eating — fresh green grass has more beta-carotene than hay does.

But why is milk overwhelmingly white? Because the tiny particles suspended in milk do not absorb light; they scatter it towards our eyes. These particles are of two types: tiny micelles, roughly one-tenth of a micron in size, composed of proteins bound together with calcium and phosphate ions; and larger fat globules, one to five microns in size. The larger particles scatter light more effectively.

In any case, the more such particles light encounters on its journey through a glass of milk, the more scattering there will be. Actually, not all colors are scattered equally. Small particles tend to scatter blue light more. In a glass of homogenized milk, there are so many fat globules and protein particles that all of the light ends up scattered towards our eyes, and we see the milk as white.

Scrutinize a glass of skim milk, and you'll notice that it has a bluish tinge. This is because the fat globules have been removed, leaving fewer particles in suspension to scatter light.

The particles it contains are smaller, so what we see is the effect of shorter-wavelength blue light being scattered more easily. In fact, if you shine a beam of light on the glass of milk and view it from the side, you'll see a blue color, but, if you place a screen behind the milk, you'll notice that the light that goes right through the glass is pinkish in color. The blue light has been scattered to the side, but the longer red waves have been transmitted through the milk.

Fog lights are amber, exploiting the fact that light with the blue color removed is less likely to be scattered. The water droplets in fog scatter the blue components of light easily, and a lot of this is reflected back to the viewer. Amber glass filters out the blue wavelengths, leaving the more penetrating longer wavelengths, which then reflect off objects, enabling us to see them.

154. How did Admiral Nelson return to England after the Battle of Trafalgar?

Aboard the *HMS Victory*, in a barrel of brandy. Nelson was killed in 1805, during the Battle of Trafalgar, but his men didn't want to feed him to the fish. Returning the body to England, however, presented them with a challenge until it occurred to one seaman that they could pickle the admiral.

Everyone knew about alcohol's preservative properties, and there was plenty of brandy on board. They immersed Nelson in the liquid (not in rum, as some romanticized accounts suggest), which they replaced with wine when the ship docked in Gibraltar. There is no truth to the rumor that Nelson's cask arrived in England partly empty because his men weren't put off by a little "admiral flavor." Nevertheless, British sailors

still use the expression "tapping the admiral" to describe an unauthorized drink.

The most celebrated naval officer in British history was buried in a crypt in St. Paul's Cathedral, and to this day his likeness stands guard 170 feet above Trafalgar Square from atop Nelson's Column.

Nelson was not the only famous military leader to be preserved in an unusual fashion. Alexander the Great's body was preserved in a large crock of honey. How did that work? Honey is a concentrated solution of sugars in water, mainly fructose and glucose, with smaller amounts of sucrose. Various other compounds, in small quantities, are responsible for the flavor and aroma. The preservative action of sugar is due to its ability to remove water from microorganisms through osmosis. It works like this: if the concentration of a dissolved material is higher outside the microbe than it is inside, then water will diffuse through the cell membrane to the outside, thereby dehydrating and killing the microbe (see p. 146). The ancient Romans already understood this phenomenon — they preserved fruits and meats by immersing them in honey. They also discovered that wounds treated with honey healed more efficiently, so Roman soldiers carried honey with them into battle. Modern research shows that honey contains compounds — other than sugars — that could have antibacterial activity. Preliminary studies have demonstrated that, at least in the laboratory, honey kills both the *Helicobacter pylori* and the *Clostridium difficile* bacteria.

Researchers still don't know whether this can happen in the digestive tract, but they are studying subjects with infections who are taking two spoonfuls of honey several times a day. It may be that, not only does a spoonful of honey help the medicine go down, but it also acts as a medicine itself.

155. Why does coffee heated in the microwave foam up when you add sugar?

It's due to a phenomenon called superheating. But to understand it, we must first understand what boiling is all about.

At the surface of a liquid, molecules evaporate continuously. If we leave a glass of water uncovered, then the liquid will eventually disappear. If we heat the liquid, then its molecules will move faster and become more energetic, and more molecules will enter the vapor phase. As a consequence, the liquid evaporates more quickly. At the boiling point, molecules all over the liquid, not just at the surface, are energetic enough to enter the vapor phase. They do this most readily by evaporating into air spaces that exist in the container. All containers have imperfections that trap air when a liquid is introduced. As these air pockets fill with vapor, they expand and begin to rise. That's why we see streams of bubbles originating from the sides or the bottom of the container.

A microwave oven heats the liquid, not its container. This means that the liquid in contact with the container is actually cooler than the liquid in the center, because heat is transferred to the walls of the container. Sometimes, the temperature difference can be as much as fifty degrees. But the liquid in the center cannot boil, because there are no air bubbles for it to evaporate into. By the time the liquid near the edge of the container reaches the boiling point, the liquid in the middle is considerably hotter. It is "superheated." If you add sugar or a teabag now, you will spur vigorous boiling, because the surface imperfections introduce trapped air bubbles into which the superheated liquid vaporizes. You can prevent accidents by placing a plastic spoon in the mug or glass while it heats to provide a surface on which bubbles can form.

156. What keeps Krazy Glue from sticking to the inside of the tube?

———

Krazy Glue doesn't stick to the inside of the tube for the simple reason that the inside of the tube is not Krazy Glue. Let's think for a minute about what this remarkable substance — which only develops its adhesive properties when activated by moisture in the air — has to do. It has to wet two surfaces and then set into a solid matrix. This hardening is usually accomplished by a polymerization reaction.

Inside the tube, Krazy Glue is comprised of simple molecules of ethyl-2-cyanoacrylate. These contain a highly reactive double bond. In the presence of hydroxyl ions, the double bonds open up and reorganize themselves to bind the small molecules into long chains. This is a polymerization. Water always contains small amounts of hydroxyl ions, enough to start the reaction, and all surfaces are covered with moisture to some degree. So, as soon as you squeeze a little ethyl-2-cyanoacrylate out of the tube, the reaction begins. The polymer is a tough, hard material, which then binds the surfaces together. The reaction goes so fast that, if you're not careful, you'll glue your fingers together. If that happens, use hot water or acetone to remove the glue.

Some physicians take advantage of Krazy Glue's ability to adhere strongly to the skin — they use a professional version of the stuff, called Dermabond, instead of sutures to close surgical wounds. In fact, you can use Krazy Glue at home to close superficial cuts when nothing else is available.

157. Why does dropping a steel wool soap pad into sudsy water destroy the suds?

Water is a good cleaning agent, but its efficacy is dramatically improved by the addition of surfactants. These substances improve water's spreading ability and help it to remove greasy materials. All surfactants share one structural feature. They are molecules with two distinct parts: a hydrophobic, or water-hating, part; and a hydrophilic, or water-loving, part. The hydrophobic part is a chain of carbon atoms from eight to eighteen carbons long; the hydrophilic part is made up of atoms, or groupings of atoms, that have an affinity for water.

This affinity is usually due to the presence of either a negative charge — as in anionic surfactants — or a positive charge — as in cationic surfactants, on one end of the molecule. While both

kinds of surfactants are effective, they have specialized uses and
are compatible with different substances. They are not compatible with each other. When present in the same solution, they
interact and neutralize one another. The positive charge attracts
the negative, forming a complex that no longer has detergent
or foaming activity.

You'll notice this when you dip a soap pad into water containing dishwashing detergent. Such detergents are anionics,
and, as the manufacturer of SOS pads confirms, steel wool pads
infused with soap use cationic surfactants. Why they use cat-
ionics, which are generally less effective cleaners, isn't clear, but
it likely has to do with steel wool's absorption capabilities.

The moral of this little story is that using steel wool in combination with dishwashing detergent won't enhance the sparkle
of your dishes. In fact, loss of suds is detrimental. Better to let
the SOS save your pans alone. Indeed, that brand name has an
interesting history. In 1917, Ed Cox of San Francisco invented
a presoaped pad for cleaning pots. His wife dubbed it SOS, for
"Save Our Saucepans."

158. How can you prevent brown sugar from lumping?

————

Nobody likes to get into a knife fight with solidified brown
sugar. Best to avoid that situation altogether, but in order to do
so we have to examine why the stuff clumps in the first place.

Think of what happens when you spill a sugary liquid, like
a soft drink, on your hands. Real stickiness only develops once
the water evaporates. Sugar is a natural glue, but its molecules
have to get very close to a surface before they can bond to it.
When dissolved in water, sugar flows easily over a surface; then,
when the water evaporates, a sticky sugar layer remains.

Sugar producers start the sugar-making process by boiling sugarcane or sugar beets in water. Then they filter the solution, concentrate it through evaporation, and allow it to crystallize. The liquid left behind, molasses, still contains a lot of dissolved sugar as well as other compounds from the plant. Purification eventually yields white crystals, but, if the producer leaves a thin layer of molasses on the crystals, the result is brown sugar. Essentially, then, the sugar crystals are coated with a sugar solution that converts to glue as the water evaporates.

The way to prevent clumping, therefore, is to prevent the moisture from evaporating. The best method of doing this is to store the sugar in an airtight container. If, however, you find yourself with some clumpy brown sugar on your hands, you'll have to restore its water content. One simple means of doing this is to put a couple of apple slices into the container. These will humidify the air and reduce the stickiness. Or you could prevent the sugar from drying out in the first place by soaking a small brick chip — or a commercially available ceramic disk, specially made for this purpose — and placing it in the container.

And why even contend with brown sugar at all? A lot of people assume that it's healthier than white, but it's not. The only reason to choose brown sugar is for the taste — and I have to admit that it tastes great on oatmeal.

159. Why do we use hot water to wash clothes?

————

Because heat causes molecules to move faster. In fact, temperature is just a measure of how quickly molecules move about.

When a substance dissolves in water, its components, be they ions or molecules, form stronger attractions to water molecules than to each other. They pull apart and are surrounded by water molecules. In other words, the water molecules wedge themselves between the units of the solute to bring it into solution. The faster the water molecules move, the greater their kinetic energy, and the greater the chance that they will blast themselves into the solute. For this reason, hot water removes water-soluble stains more quickly. But not all stains are water soluble.

The molecules that make up oily stains have a stronger affinity for each other than for water. Water molecules cannot wedge themselves into the network of fatty molecules — they are repelled. Oil floats on water, and oily stains do not dissolve. So we use detergents. These are molecules with one end that has an affinity for oil and one end that has an affinity for water. The oil-soluble end anchors itself in the stain; the other end remains loose in the water (see p. 138). As the water molecules move around, they tug at the detergent, which in turn tugs at the fatty molecules in the stain, breaking the stain apart. At higher temperatures, this molecular dance occurs at a greater speed and with greater energy, removing the stain more readily.

But when it comes to a detergent to which manufacturers have added enzymes to break down protein stains, a high water temperature is actually detrimental. The enzymes degrade at high temperatures, and that's why enzyme detergent manufacturers advise consumers to use lukewarm water with their products.

The relation between an increase in solubility and an increase in temperature does not always hold, however. Gases are less soluble at higher temperatures, because gas molecules in solution have a very weak attraction for the solvent. Since, at a higher temperature, the gas molecules are moving more rapidly, they escape more easily from the surface of the solution. A cold carbonated beverage will dissolve more carbon dioxide than a warm one. Just heat a soda and watch the bubbles evolve. The same thing happens when you let a glass of cold water stand — the dissolved air will be released in the form of bubbles.

There is a limit to how much the solubility of gases in water will increase as the temperature drops. When water freezes, its molecules take on a crystal pattern and squeeze out the molecules of a solute. That's why it's not such a great idea to put a bottle of beer in the freezer. Inside the bottle, a lot of carbon dioxide has been dissolved under pressure, and as the liquid freezes the gas comes out of solution. This could lead to an explosion, and you'll wind up covered with beer stains. But you know what to do — wash your clothes in hot water and detergent.

160. What is a poison?

This is a trick question, because virtually any substance you can ingest, inject, inhale, or absorb through your skin can be a poison if the dose is high enough. Of course, a high dose can sometimes be very small — it all depends on what the substance is. The ancient Chinese occasionally carried out executions by forcing condemned individuals to eat salt until they died. It took a lot of salt. By contrast, just a few micrograms of botulin toxin can kill a man.

How is it that minute quantities of a substance can cause such devastation? Because poisons interfere with critical processes in the body. Different poisons have different modes of action. Socrates was sentenced to die "by the cup" in 402 B.C. for corrupting the minds of Athenian youth. The cup contained the juice of the hemlock plant, which the ancient Greeks often used for executions. Coniine, the active ingredient in hemlock, kills by interfering with the activity of the nervous system.

Symptoms usually start in the feet. There is a sensation of cold-ness and creeping paralysis. The legs get numb, then the hands get cold, and finally the muscles needed for breathing become paralyzed. Death is quick.

Some poisons interfere with the body's enzymes. Cyanide, for example, inactivates an enzyme called cytochrome, which our cells need to use oxygen. Death results from lack of oxy-gen, even though there is plenty of oxygen around. A lethal dose can be as small as fifty milligrams, which may sound like a tiny amount, but it contains ten billion, billion molecules! Other poisons have very different modes of action. Carbon monoxide can also cause oxygen starvation but by a different mechanism — it displaces oxygen from hemoglobin molecules, which deliver oxygen to cells. Some snake venoms interfere with blood coagulation, causing the victim to bleed to death.

Recently, a surprising study reported in the prestigious jour-nal *Nature* cast doubt upon the adage "Only the dose makes the poison." After analyzing a large amount of animal data, researchers concluded that at certain doses dioxins have protec-tive effects. Dioxins are notorious compounds. Many consider them the most toxic ever created by humankind. Amazingly, at a specific (very small) dose, animals treated with dioxin had a lower risk of developing tumors than they did at a higher or lower dose.

This phenomenon is far from uncommon. Of the thousands of chemicals that researchers have examined in many animal spe-cies, toxicity dips before rising sharply. Perhaps, in appropriate small quantities, chemicals exercise cells' defenses, making the cells more resistant to insult. Or maybe low levels of the chemi-cals that cause mutations in DNA activate the enzymes that repair previous DNA damage. Poisons are complicated indeed.

161. Why is glucose added to salt?

Check the label on a box of salt, and chances are you'll find glucose — or, as it's sometimes called, dextrose — listed as an ingredient. The reason it's there has to do with reducing the risk of thyroid problems.

The thyroid gland, located in the neck, produces hormones that control the rate of energy production in all our cells; it therefore influences the way all our organs function. Insufficient production of thyroid hormones can cause weight gain, lethargy, constipation, clammy skin, and perhaps even hair loss and premature graying. Lack of thyroid hormones in utero can cause cretinism. Various factors can cause the thyroid gland to be underactive, including a lack of iodide in the diet. Thyroid hormones have iodine in their molecular structure, and iodide (a negatively charged ion) is the form of the element that we require in our diets in order to synthesize thyroid hormone. If we don't consume enough iodide, then our thyroid glands cannot make enough hormone. The gland will struggle to withdraw the available iodide from the blood, and as it does so the gland enlarges, producing goiter, a typical sign of iodine deficiency.

Until the early part of the past century, the American Midwest was known as the Goiter Belt, because the region's soil lacked iodine. So, the crops grown there, and the animals that fed on them, contained little iodine. Scientists wondered why goiter was endemic in the Midwest but almost never appeared in coastal regions. By 1920, the iodine connection had become apparent. Seawater is rich in iodine, and coastal soils are therefore rich in it as well. Authorities throughout the U.S. worked to remedy the situation. In 1924, the state of Michigan began to experiment with adding sodium iodide to salt; the city of Rochester, New York, added it to the drinking water.

Today, salt with added potassium iodide is common. But the solution to the goiter problem wasn't as simple as that. What complicated things was the fact that iodide slowly converts to iodine in moist air. Since iodine is volatile, salt that just has potassium iodide added to it slowly loses its power to protect against goiter. That's why, after much experimentation, manufacturers started adding iodine stabilizers to iodized salt. Iodine stabilizers convert iodine back to iodide.

At one time, they added sodium thiosulfate, but it sounded too "chemical," and people became worried. So, the manufacturers switched to dextrose, which is as effective and sounds innocuous. Sometimes, they add sodium bicarbonate, because the oxidation of iodide occurs readily in an acid solution and not in a base; bicarbonate produces basic conditions. And occasionally they might use disodium phosphate or sodium pyrophosphate to provide the alkaline conditions. These are also sequestering agents, which bind trace metals that catalyze the oxidation of iodide to iodine.

Unfortunately, these simple protective measures are not implemented everywhere in the world. In India, an estimated 250,000,000 people suffer from iodine deficiency and the accompanying lethargy and decrease in motor skills and mental acuity. About nine million babies are born with cretinism annually. Salt is often sold in big chunks, because in this form it resists humidity better. The only way to iodize the salt chunks is to spray them with a solution of potassium iodate. This, though, makes the large salt crystals look dirty, so people often wash them off before crushing them. They don't realize that cleanliness is not always a virtue.

162. How do smoke detectors work?

Fires kill more people in North America than all natural disasters combined. Many die when they are overcome by smoke without warning. Smoke detectors can provide us with that warning and the time we need to escape.

There are two types of detector: the photoelectric detector and the ion-chamber detector. The first works on the principle that smoke particles scatter light. Imagine a sunbeam shining into a darkened room. Its path is made visible by light reflected from dust particles. Inside the photoelectric detector, a tiny beam shines from a light-emitting diode across a little chamber. A light detector is located in another part of the chamber, out of the beam's path. But when smoke particles are present, they scatter the light towards the detector. The light hitting the detector triggers an electronic circuit, which activates the alarm.

The ion-chamber variety is the more popular type of detector. It employs a small amount of a radioactive material — americium-241 (half-life 458 years) — to generate an electrical current. Am-241 is made by neutron bombardment of Pu-239. It radiates charged particles, called alpha particles, which bombard air molecules inside the detector, knocking off electrons to generate ions. These ions complete an electrical circuit. The presence of smoke particles or gases reduces the mobility of these ions, thus reducing the electrical current. The reduction in current sets off the alarm.

Ion-chamber detectors are very useful because gaseous combustion products that form even before there is any smoke can set them off. But this also means that shower steam can trigger them, and that can be a nuisance. However, the protection the detector offers eclipses such minor annoyances. Of course, that protection is only available if the detector is working properly. Remember that pushing the test button on the unit will only test the alarm circuitry. You should test your detector every couple of weeks by blowing out a candle flame just beneath it. And keep in mind that the ion-chamber detector has a limited life span — about ten years — because the americium loses its activity.

Some people worry that the radioactive substance in these detectors poses a risk, but the quantity of americium they contain is very small, and it's well shielded. Furthermore, since alpha particles travel only about five centimeters from their source, americium would only be dangerous if the source were eaten. So, don't eat smoke detectors. And, above all, remember that even the best smoke detector is useless if its battery doesn't work. Researchers report that about 25 percent of detectors installed in homes are nonfunctional, and this is mainly because the battery is dead or someone has removed it to power a toy. If you do that, you could be playing with your life.

163. Why do the water bottles used by bicyclists have more of a plastic smell in warm weather?

Almost all such bottles are made of polyethylene, a common plastic. The word *poly* refers to a molecule made up of many repeating units. Polyethylene is made by chemically combining many small molecules of ethylene. Water bottle manufacturers add nothing to the polyethylene, not even ultraviolet light stabilizers, which makers of items such as garden furniture use to keep the plastic from breaking down through exposure to sunlight. The problem is these stabilizers leach into water, so they aren't safe for water bottles. As a consequence, water bottles will eventually crack from sun exposure.

So, if the polyethylene contains no additives, what smells? Actually, it's the plastic itself. In order for us to pick up any smell at all, receptors in our noses must be engaged by molecules. Generally, polymers are not volatile, so they don't smell. But when ethylene molecules are joined together to make polyethylene, some short-chain compounds are produced as well. These are more volatile, and they do have a smell. Why more so in hot weather? Substances enter a vapor phase more readily with heat. Just compare the aroma of a freshly baked loaf of bread with that of a cold loaf.

Should we be concerned about that plastic smell? No — even short chains of polyethylene are not toxic. But if the smell bothers you, then wash out the bottle a few times with boiling water — that will remove the volatile compounds. Then you can go bicycling in the desert without having to drink from an odiferous bottle.

164. A coffeepot comes with the following instructions: "Clean every week with vinegar if used only to boil water, clean every other week if used to make coffee." Why?

The instructions don't seem to make sense. How could a machine that's flooded daily with a dark, pungent liquid require less frequent cleaning than one that has to deal only with nice clean water? The reason that kettles, coffeemakers, and, for that matter, boilers require frequent cleaning has to do with the minerals dissolved in the water.

The culprit is a soluble compound of calcium called calcium hydrogen carbonate. When we heat water, this substance decomposes to yield carbon dioxide, which, being a gas, dissipates into the air, and calcium carbonate, a water-insoluble compound that forms a deposit called scale. And how does the calcium hydrogen carbonate get into the water in the first place? It comes from limestone deposits in the ground.

Chemically speaking, limestone is calcium carbonate. Although calcium carbonate is insoluble in water, it does dissolve in a weakly acidic solution. Surface water is weakly acidic, because carbon dioxide in the air dissolves in rain and reacts with the water to form carbonic acid. As the acidic surface water percolates through the ground, it converts insoluble calcium carbonate to soluble calcium hydrogen carbonate. When we heat water, we reverse this reaction, causing deposits of what amounts to limestone to build.

If calcium carbonate dissolves in weakly acidic rainwater, it stands to reason that it dissolves even more readily in a stronger acid, such as vinegar. This, then, explains why the manufacturer of that coffeemaker recommends using vinegar to clean the unit. But why do we need to do this less often if we're using the machine exclusively to brew coffee? Simple. Coffee is acidic. Every time we brew a batch, some of the scale inside the pot dissolves.

There is no reason to worry about calcium carbonate dissolving in your coffee. Calcium carbonate is actually the active ingredient in many antacid preparations. Just as it neutralizes excess stomach acidity, it neutralizes excess coffee acidity. So, finally, it all makes sense. Since coffee is more acidic than water, it is better at preventing the buildup of limestone scale. But what the brown brew may do to our internal plumbing is a different story altogether.

165. What is processed cheese?

Processed cheese was invented in 1916 by J. L. Kraft, a cheese merchant who was plagued with complaints that his product was not of a consistent quality. Kraft hatched a scheme: he would mix a variety of cheeses and blend them with water to produce a uniform product. But to ensure that his product had a smooth consistency, he'd have to devise a method of preventing the fat and the water from separating. Sodium monohydrogen phosphate, Kraft discovered, was an ideal emulsifier. Today, processed cheese makers use essentially the same procedure.

Incidentally, Kraft's familiar yellow slices are yellow for the simple reason that consumers think they should be; the cheese is dyed with carotenoids extracted from a variety of plants. Nutritionally, processed cheese is roughly equivalent to regular Cheddar cheese.

166. How can vampires save human lives?

Count Dracula often entered the bedrooms of his victims in the form of a bat. Why a bat? Because folklore has it that bats are the souls of the departed. But let's leave the problem of vampires to vampirologists and the mysteries of the soul to religion scholars.

There is no doubt, however, that vampire bats exist. Most bats feast on insects, and they are excellent pest eradicators, but vampire bats tend to be pests themselves. These tropical creatures feed at night, lapping up the blood that flows from small incisions they make in their prey with their sharp teeth. Usually, the victim — a cow, say — isn't even disturbed by the bite. On occasion, a vampire bat will prey on a human, perhaps biting an exposed big toe. Aside from the fact that such a bite could give the victim rabies, it's no big deal.

The vampire bat's metabolic rate is very high, so the creature requires a lot of blood; and, due to its high-liquid diet, it drinks and urinates at the same time. In order to keep the blood flowing from the wound, the vampire bat has a protein in its saliva that prevents coagulation and clot formation. This protein is called bat plasminogen activator, or bat-PA. Many are already familiar with this type of substance, because physicians prescribe human tissue-type activator (t-PA) to people who have suffered a heart attack. The idea is that the activator minimizes damage to the heart by dissolving the clot responsible for cutting off the blood flow through the coronary arteries.

One of the problems with this treatment is that it not only dissolves existing blood clots but also prevents clot formation. This translates into an increased risk of internal bleeding. It seems that bat-PA may be superior to the currently used substance, t-PA, because it is capable of dissolving existing clots without impairing the blood's ability to coagulate if necessary. So, who knows? Someday, hospitals may stock vampire bats. But they'll have to keep them away from the blood bank.

167. Why does the juice of the grated potatoes in Christmas pudding turn black when you add baking soda?

The chemistry of the darkening of grated potatoes is actually quite complex. It involves two different reactions, and these occur to different extents.

First, there is the browning reaction, which we see with apples. Damage to cells releases an enzyme, polyphenoloxidase, which causes compounds called phenols to combine with oxygen to form melanin-like pigments. In the case of potatoes, the prime culprit is chlorogenic acid. The extent to which the browning reaction occurs depends upon the degree of acidity, or pH. Polyphenoloxidase is most active around pH 7, and it's less active under acidic conditions. Potatoes contain naturally occurring acids that slow down the browning reaction. But if we add baking soda, a base, to potatoes, then the pH rises, and the browning reaction proceeds more quickly.

The second reaction has nothing to do with polyphenol-oxidase. Potatoes contain a small amount of iron, which they absorb from the soil in an ionic form known as ferrous iron. Chlorogenic acid forms a colorless chemical complex with ferrous ion. When a potato is cut or cooked, the ferrous ion reacts with oxygen and converts to a form called ferric iron. The ferric-iron/chlorogenic-acid complex is black, and it causes the blue-black spots you often see in cooked potatoes.

The solution to the darkening problem triggered by either reaction is quite simple: ascorbic acid, or vitamin C. It inactivates polyphenoloxidase, and it also forms a complex with iron, preventing formation of the ferric-iron/chlorogenic-acid complex. Furthermore, it acts as an antioxidant, reducing ferric iron to ferrous, and thus it also destroys color. Plus, it's good for you.

In any case, the darkening reaction doesn't pose a health hazard. Even if your potatoes have turned black, you can still eat them. After all, potatoes are a true comfort food. Due to their high carbohydrate content, they elevate serotonin levels in the brain, which has a calming effect. And, believe it or not, researchers have discovered that potatoes harbor small amounts of naturally occurring benzodiazepines, compounds in the same chemical family as Valium. So, go ahead and enjoy your Christmas pudding, white or black.

168. Why don't your eyes water when you cut cooked onion?

——

Onion chemistry is complex and fascinating. We have been intrigued by this vegetable ever since our prehistoric ancestors gathered and cooked wild onions. By the time of the First Egyptian Dynasty, five thousand years ago, onions were widely consumed, prized not only for their flavor but also for their supposed medicinal properties.

At various points in history, people have credited onions with preventing colds, loosening phlegm, correcting indigestion, eliminating parasites from the digestive tract, stimulating appetite, disinfecting wounds, and inducing sleep. To ancient peoples, the onion's concentric circles symbolized eternity. For this reason, onion-shaped domes became popular in Eastern European architecture — the idea was that the buildings they adorned would stand forever.

Onions may not make us live forever, but some of their components may indeed have medical benefits — lowering cholesterol and blood pressure and perhaps even reducing the risk of cancer. This is why their chemistry has drawn a great deal of attention. We now know that slicing an onion sets off a cascade of reactions that begins with alliin, a compound that occurs naturally in onions and garlic. When the tissues of the onion are disturbed, allicinase — an enzyme that converts alliin to allicin — is released. This, in turn, breaks down to syn-propanethial-s-oxide, the stuff that makes your eyes water.

Frying the onion causes yet another reaction, resulting in the formation of bispropenyl disulfide, which has a sweet smell and a sweet taste. Dozens of other chemical reactions also take place upon the application of heat, possibly forming other sweet substances and causing the destruction and evaporation of the strong-tasting and strong-smelling components. The reason cooked onions don't make us cry is that syn-propanethial-s-oxide has evaporated.

But the goal of onion research is more than just culinary. For example, researchers in West Germany have discovered that compounds in onions called thiosulfinates can relieve asthma. And Bela Karoly, the renowned former Olympic gymnastics coach, swears by the old Transylvanian remedy of applying a cooked onion to an inflamed joint.

169. A newsletter written by a physician suggests that you can lose a significant amount of weight by chomping on ice cubes. Why should this doctor be sent back to school?

So that he can learn the difference between "small calories" and food calories. "Head for the Freezer to Lose Weight!" screamed a headline in the physician's newsletter. And this newsletter was no tabloid — it was produced by a trained medical professional and had a huge circulation. The doctor was claiming to have made the astounding discovery that, if a person swallows ice, then his or her body will expend some energy to melt it and more energy to bring the resulting water to body temperature. He calculated that it takes about 240 calories to melt two quarts of ice and raise the water's temperature to 37°C (98.6°F).

What an easy way to lose those unwanted pounds! All you have to do is munch a couple of quarts of ice per day and watch the weight melt away. The doctor provided testimonials to bolster his discovery. An Oklahoma City physician maintained that he'd dropped thirty-five pounds using the ice method. People laughed at him, he said, until he lost the weight.

Our astute newsletter writer recommended to his readership that they crush the ice before munching it to avoid breaking their teeth. He also advised that they swallow the ice in small quantities to ward off brain freeze — the sharp headache that often comes from eating cold foods. Actually, our weight-loss hero should have heeded his own advice — he must have suffered from brain freeze himself, because he really got things muddled. Did he forget what he'd learned back in chemistry class? A calorie is a unit of measure; it's the amount of heat needed to raise the temperature of one gram of water by one degree Celsius. In this sense, the doctor's calculation is indeed correct. But — and there is an enormous "but" here — this definition does not apply to food calories. It applies to so-called small calories.

A food calorie is equivalent to a thousand small calories. So, the doctor's entire weight-loss scheme was based on one big error. You could chomp on an iceberg all day long and never lose an ounce. If you're serious about losing weight, you know what you have to do: eat less and exercise more.

170. Why do boiled eggs sometimes smell like old sneakers when you remove their shells?

Eggs are amazing. They undergo all kinds of chemical reactions. Put them on a shelf for a while, and they rot. Put them in boiling water, and they harden. Put them under a chicken for thirty days, and they yield chicks. If the eggs were fertilized, that is. Eggs can also produce some pretty impressive smells. Rotten eggs smell because they contain hydrogen sulfide, a chemical that closely resembles the one that gives sweaty feet that have been laced into sneakers all day their pungent aroma. In this case, the culprit is methyl mercaptan, CH_3SH, a close relative of hydrogen sulfide, H_2S.

Eggs produce hydrogen sulfide when their proteins, which contain both sulfur and hydrogen, break down. The liberated sulfur and hydrogen combine to produce hydrogen sulfide, one of the smelliest existing compounds. But it's not the smelliest. That honor may go to methyl mercaptan, produced by bacteria on our skin when those bacteria find themselves in a moist, airless environment. So, that's why rotten eggs and feet have similar smells. But now to the boiled eggs.

As an egg cooks, it releases hydrogen sulfide. Although the hydrogen sulfide smell is unpleasant, at such low concentrations it's not dangerous. But at high doses, it is deadly — workers have been killed by hydrogen sulfide after falling into manure pits, where bacteria convert sulfates to hydrogen sulfide. Don't worry about it when you're cooking eggs, however. If you boil an egg in its shell, the gas does not escape until you peel off the shell — hence the smell. If you fry the egg instead, the hydrogen sulfide is released as soon as it forms, so there's no gas buildup and little smell.

Although hydrogen sulfide conjures up memories of old sneakers, it can also trigger other recollections. Remember stink bombs? They share this interesting bit of chemistry with eggs and sneakers. And what about the expression "Last one in is a rotten egg!"? It doesn't really apply to egg cookery, because the first egg in the water will be heated for a longer time and is likely to produce more hydrogen sulfide than the last one in, providing you remove them at the same time.

171. What is camphor?

It wasn't so long ago that mothers hung little bags of camphor around their children's necks when they had colds. The strong, medicinal smell probably had some effect in opening up congested nasal passages, but its real value as a treatment was likely that it prevented the virus from spreading by keeping others at a distance.

Camphor comes from the camphor tree, which grows in the Orient, and it's isolated by the process of steam distillation. Today, however, most camphor is made synthetically from pinene, extracted from pine trees. Its effectiveness as a decongestant is questionable, and problems associated with its use are well documented. Skin irritation, headaches, dizziness, confusion, and even hallucinations have been linked with inappropriate applications of camphor.

What do I mean by "inappropriate"? How about rubbing camphorated cream on your partner before hitting the dance floor? Read the label on any camphorated cold remedy, and you won't see this listed as a recommended application for the product. Yet this is exactly what dance-loving teenagers in the north of England started to do a few years ago. Why? Because they'd heard that camphor prolonged the effects of the designer drug Ecstasy, a dangerous substance that's all the rage among those who attend those strange, ritual-like gatherings called raves.

In an attempt to curb drug abuse, British police had cracked down on the rave scene, and they had started searching kids as they entered the dance venues. The ravers couldn't just take their Ecstasy at home before the event, because its effects are relatively short-lived, so they carried the drug with them. Then some clever closet chemist figured out that, if you smeared yourself with camphor, then the effects of the Ecstasy would last longer — you could ingest the drug earlier in the evening and thus avoid getting caught red-handed with it at the rave.

As this strange story began to spread, the Ecstasy connection was lost. Before long, ravers were telling each other that the camphor rub alone had hallucinogenic effects. Its effects were likely just psychological, since the amount of camphor absorbed through the skin is relatively small. But a high dose, if ingested, could certainly act as a stimulant. There have been cases of children suffering seizures from drinking camphorated oils sold as body rubs.

Camphor is also available as tiger balm, a traditional Asian remedy. The balm has surfaced at raves, giving parents and police headaches. Someone should tell them that tiger balm can actually cure a headache. A study conducted at Australia's Monash University compared the effectiveness of tiger balm to acetaminophen — the active ingredient in products like Tylenol — and to a placebo. The study's sixty subjects either took acetaminophen pills or rubbed a small amount of ointment on their temples three times at half-hour intervals. They also recorded the severity of their headaches on a scale of one to seven over a three-hour period. To the surprise of the researchers, the tiger balm was as effective in treating tension headaches as the acetaminophen. Both were significantly better than placebo. There seems to be no simple explanation for why this treatment works, although the balm may relax the muscles in the forehead.

What if you don't suffer from tension headaches and you aren't into raves? Well, then, you can always use camphor as a moth repellent. And that's the rubdown on camphor.

172. What did Moses have to do with anthrax?

The Pharaoh Ramses had already seen Moses's rod turn into a serpent, the River Nile run red with blood, and the land of Egypt overrun with frogs and lice. Yet his heart was hardened, and he would not let the Hebrews go. The Lord then instructed

Moses to throw dust into the air, and as the dust spread all who came in contact with it, human and beast alike, broke out in boils. Could this plague possibly have a scientific explanation?

One of the oldest recorded human and animal diseases is anthrax. Today, we know that it's caused by a bacteria, *Bacillus anthracis*, which is found in soil and can retain its virulence for years. Throughout history, anthrax outbreaks have wiped out herds of grass-eating animals. Humans have generally contracted the infection by handling the wool, hides, or bones of infected animals. Pus-oozing boils are a common symptom, and they can lead to blood poisoning and death. Woolsorter's disease is a particularly dangerous form of anthrax — it attacks the lungs, and it's often fatal.

Anthrax has undoubtedly caused a great deal of misery, but it was instrumental in the development of highly effective forms of treatment for infectious diseases. The ancient Chinese had actually discovered that they could successfully treat boils with moldy curds or soybeans, but it wasn't until some two thousand years later that scientists gained real insight into the nature of infectious diseases. This came about when the French sugar industry asked Louis Pasteur to find out why beet-sugar mash sometimes turned into a useless goo instead of fermenting into alcohol. The brilliant chemist investigated the problem and learned that microorganisms in the air were responsible. He also discovered that at high altitudes — like in the Alps — where the organisms were absent, fermentation did not occur.

Pasteur went on to suggest that human and animal diseases might also be caused by such microorganisms, and he formulated the "germ" theory of disease. To prove his theory, he showed that anthrax was caused by a bacterium that was transferred from animal to animal. But even more exciting was his observation that animals injected with a weakened form of the bacteria developed an immunity to the most active varieties of the organism. Pasteur organized a public demonstration in which he vaccinated part of a herd of sheep against anthrax and then exposed the whole herd to live bacteria. The vaccinated animals did not contract the disease. Today, such vaccines are widely used to prevent anthrax outbreaks.

But the story doesn't end there. Pasteur noted that, when anthrax bacteria were inoculated into sterilized urine, they multiplied rapidly, but, when other bacteria were also present in the urine, the microorganisms were killed. This was a clear demonstration of the destruction of one living thing by another — the very first observation of an antibiotic in action.

173. How did sick silkworms and an accidental spill lead to the invention of rayon?

———

Once more, we look to Louis Pasteur. In 1862, French silk industry representatives prevailed upon him to save them from a potentially catastrophic epidemic that was afflicting silkworms. Pasteur traced the problem to a tiny parasite that infected the worms and the mulberry leaves upon which they fed. Upon his recommendation, the worms and their food supply were destroyed. After getting off to a fresh start with healthy new worms, the silk industry flourished once more. Through a peculiar sequence of events, this scientific rescue led to the development of silk's first real commercial competitor: rayon.

Count Hilaire de Chardonnet was an assistant to Pasteur during the silkworm saga. He became fascinated with the way silkworms spin silk, and he thought that, by understanding what the worm did, humankind could also learn to spin the desirable fiber. He was never able to duplicate silk, but he did manage to produce the first commercial artificial silk. And he discovered the method by accident.

Back in those days, photographers used a substance called collodion to coat and protect their photographic plates. Louis Menard had developed collodion in 1846 by dissolving nitro-cellulose (cotton treated with a mix of nitric acid and sulfuric acid) in alcohol. The transparent, gelatinous liquid dried to a hard, colorless film, and it was also popular as a dressing for cuts and burns. One day, Chardonnet was working with his photographic plates when he accidentally spilled a bottle of collodion. He left the spill for a while, and, when he returned with a cloth to clean it up, he saw that the collodion had formed long, silk-like filaments. Due to his work with silkworms, he recognized the significance of the discovery. The independently wealthy Chardonnet spent the next six years working on his invention. He used the cellulose in mulberry leaves to make a nitrocellulose solution, which he squeezed through a shower-head-like device to produce thin filaments that could be spun into fabric.

The amazing "Chardonnet silk" was exhibited at the Paris Exhibition of 1889. Financial backers came calling, and the "silk" — named rayon, because rays of light enhanced its lustrous shine — was being produced commercially by 1891. The first rayons were highly flammable, and the factory workers who made the new material began to refer to it by a less endearing name: mother-in-law silk.

174. What antibiotic was discovered thanks to a sick chicken?

Almost everyone has heard the story of Alexander Fleming's fortuitous discovery of penicillin when one of his bacterial culture dishes became contaminated with a mold. But many other antibiotics have similarly fascinating histories. Consider, for example, streptomycin.

By the 1940s, it had become evident to scientists that many living organisms were capable of producing antibiotics. At Rutgers University in New Jersey, Dr. Selman Waksman was investigating the fungus *Streptomyces griseus*, which seemed like a good source of these drugs. As luck would have it, a poultry breeder brought a sick chicken to the university's Agricultural Research Station. A vet examined the chicken and discovered a small white spot in its throat, which turned out to be a fungus of the *Streptomyces griseus* variety. The vet knew about Waksman's research and sent over a sample.

Waksman cultured the fungus and found that it produced a substance that was effective against many disease-causing organisms. He obtained soil samples from the poultry breeder's yard and eventually isolated a drug, which he named streptomycin. In India, health authorities used streptomycin in their battle against the plague, and it reduced the death rate from 70 percent to 4 percent. In 1944, physicians at the Mayo Clinic in Minnesota successfully used streptomycin to combat tuberculosis. Although some complained of side effects such as dizziness, visual disturbances, and hearing problems, streptomycin freed people from tuberculosis sanatoriums. Later, researchers discovered that, by combining streptomycin with pantothenic acid, they could greatly reduce the incidence of side effects. Streptomycin is one of the drugs used today in the treatment of tuberculosis.

Since its discovery, pharmaceutical companies have systematically examined soil samples, often asking their employees to bring specimens home from their travels. This is how the cephalosporins, isolated from an organism living near a sewer outlet in Italy, and erythromycin, discovered in a Philippine soil sample, were added to the antibiotic armamentarium. Sometimes even dirt can have the right chemistry!

175. Why is freshly picked corn so much sweeter than corn that has been sitting around?

Corn on the cob tastes great. Especially when it's fresh. You just can't match the flavor of an ear that's been dropped in boiling water for a few minutes as soon as it's picked. If you compare its flavor to that of an ear that's been off the stalk for a while — well, there's just no comparison. And here's why.

Cells in corn manufacture glucose through photosynthesis and enzymatic reactions; then they convert glucose into starch. In living corn, some glucose is always forming, and some is always converting to starch. Within a few hours of picking, however, the chemical composition of corn changes dramatically, because glucose is no longer being produced, but the enzymatic reactions that convert it to starch continue. This makes the kernels more mealy and less sweet. Some of the sugar is also used as a source of energy to fuel the continuing reactions. As much as 40 percent of the sugar in an ear of corn can be lost within six hours of picking.

Heat inactivates enzymes, so when fresh corn is immediately plunged into boiling water, the conversion of sugar to starch is arrested. This aspect of corn chemistry presents a real problem when it comes to freezing cobs. In a frozen state, the enzymes keep working — albeit much more slowly — and this leads not only to sugar loss but also to the destruction of some of corn's flavor compounds. So, we have to blanch corn before freezing it. But this, too, presents a problem. During the blanching process, the kernels start to heat up from the outside, and, by the time the temperature near the cob is high enough to inactivate the enzymes, the outside of the kernels is absorbing water and cooking. The end product will be soggy.

Researchers at Cornell University devised a simple solution to this dilemma. They drilled lengthwise through the cob, water flowed into the narrow hole, and the kernels heated from both sides at once. The cored corn not only fared better in taste tests, but it also saved energy because it blanched faster.

176. When were coins first introduced?

Can you imagine what life would be like without coins? No gumball machines. No laundromats. No coin tricks. Of course, we've had coins for a very long time. King Croesus of Lydia probably minted the first gold coins around the sixth century B.C. We know this because, in 540 B.C., Polycrates of Samos was found guilty of counterfeiting coins.

In most parts of the world, coins were the popular currency, but in a few places people favored other items — like beetle legs or bits of tufted string made from the fur of fruit-eating bats. Even today, in some parts of Fiji, porpoise teeth are accepted as currency. There, natives hard up for money just drive porpoises into shallow waters, where the creatures are smothered by mud — cash crisis solved. Cattle have served as currency. In fact, our word *pecuniary* comes from the Latin *pecu*, for "cow." That's why coins have heads and tails.

Coin counterfeiting has a long history. No less illustrious a personage than Isaac Newton took this bull by the horns. Late

in his career, he became master of the mint, and he started to obsess about counterfeiters. He designed coinage to stymie the criminals. But it was a chemist who uncovered one of the cleverest counterfeiting schemes. In the eighteenth century, Joseph Black was professor of chemistry at Glasgow University. In the course of his work, he noticed that, when he heated calcium carbonate, it gave off a gas — carbon dioxide. He had carefully weighed the carbonate before and after heating, and the decrease in weight represented the loss of carbon dioxide.

Black's weighing skills certainly came in handy — and not just in the laboratory. It was common practice in those days for students to pay their professors directly for their instruction. The professor would install himself at the entrance to the lecture hall and collect gold coins from the students as they entered. Some enterprising students tried to reduce their fees by shaving a little gold off their coins. Black's students, however, never got away with this trick, because the canny professor weighed the coins before accepting them.

DR. JOE AND WHAT YOU DIDN'T KNOW

I think this business of giving gold coins to the professor is a good idea. With the cutbacks in education, maybe we should reinstitute it. But some clever students would probably still figure out a way to cheat.

177. Why are scientists who are experimenting with genetically modified foods interested in "daltons?"

———

Genetic modification involves the transfer of a gene from one organism to another. Genes are the segments of DNA molecules that instruct cells as to which proteins they should produce. Corn, for example, can be genetically modified with a gene from the *Bacillus thuringiensis* bacterium, which codes for the production of an insecticidal protein. Bt corn can therefore protect itself against pests, so it requires less pesticide application.

Since many allergens are proteins, a novel allergen can be introduced whenever a gene is transferred. For example, when scientists attempted to improve the nutritional quality of soybeans by modifying them with a Brazil nut gene, people who were allergic to Brazil nuts reacted to the soybeans. The soybeans in question were earmarked for animal feed, but scientists tested them for allergenicity in case they somehow entered the human food supply. As a result of this testing, the soybeans were never marketed.

Critics of genetic modification argue that other transferred allergens may not be discovered until it is too late. Actually, scientists have a pretty good handle on which proteins are most likely to produce allergic reactions, and they carefully screen genetically modified foods for them. A protein that doesn't break down easily when tested under conditions that mimic the human digestive system is a good candidate for allergenicity. Such a protein is more likely to pass through the intestinal wall into the bloodstream, where it can interact with cells of the immune system. Specific amino acid sequences in proteins that cause allergic reactions have also been identified, and these are catalogued in data banks. This allows researchers to compare them to amino acid sequences in novel proteins that are introduced into foods.

The size of protein molecules, as determined by their molecular weight, also provides us with a clue about their allergenic potential. And this is where "daltons" come in. Named after John Dalton, the British scientist who introduced the idea that elements are made up of atoms, the dalton is the fundamental unit of molecular mass (see p. 19). One dalton is one-twelfth the mass of an atom of carbon-12 (the carbon isotope that has six protons and six neutrons in its nucleus), and it is roughly equivalent to the mass of a hydrogen atom. Most allergens have molecular weights in the ten thousand to forty thousand dalton range.

The bottom line is that proteins introduced through genetic modification are more thoroughly investigated for potential allergenicity than are naturally occurring proteins in food. There is no evidence that any human has ever had an allergic reaction to a novel protein present in any genetically modified food on the market.

In 2000, StarLink corn, which had been approved for animal feed exclusively, somehow made its way into taco shells. A huge controversy erupted. The corn had not been approved for human consumption because it contained a protein — CRY9C — that, in a "simulated gastric environment," had not been readily digested. This study suggested that the corn had the potential to produce an allergic reaction. In fact, the protein contained no known allergenic sequence of amino acids and was present only to an extent of about 1 percent of the total corn protein.

Furthermore, an allergic reaction cannot occur on first exposure to a substance. It takes repeated exposure to build up the antibodies that eventually react with an allergen to produce an allergic reaction. Since the manufacturer of the suspect taco shells immediately removed the product from the market, the chance of a reaction, even if the protein did turn out to be allergenic, was minimal. The antibiotech rhetoric generated by the StarLink issue in some circles is not supported by the facts.

INDEX

Abrasion 57
Acetaminophen 223
Acetic acid 17
Acetobacter aceti 17
Acetylcholine 170
Acid or alkaline residue 106
Acid rain 67, 89, 109
Aconite 42, 159–60
Acrylamide 22–24
Acrylates 133
Addison's disease 72
Adenine 121
Agnuside 46
Agutter, Paul 170
Alcohol 64
Algae 68, 185
Allergenicity 231, 232
Allergens 231
Allergy 81
Alexander the Great 200
Allicin 218
Allicinase 218
Alliin 218
Allyl isothiocyanate 84
Alpha particles 211
Aluminium 93–94, 140–41
Americium-241 211–12
Ammonia 28, 158–59

Ammonium bicarbonate 28
Ammonium carbamate 28
Ammonium isocyanate 117
Ammonium perchlorate 93
Angel's trumpet 170
Aniline 161
Annatto 59
Antifog sprays 191
Anthrax 224–25
Ants 66
Artificial snow 192
Asklepios 14
Aspergillus 193
Astaxanthin 58
Atropine 41, 170, 171

Bacampicillin 80
Bacillus thuringiensis 231
Bailey, William J. A. 148
Baking soda 193–94
Bat plasminogen activator 216
Balard, Antoine-Jérôme 112–14
Bald-Headed Men of America 38
Baldness 38–39
Balloon 98–99
Banting, Dr. Frederik 157–58
Barbie 31–33
Barium sulfate 88

Barium sulfide 88
Battery 28
Bauxite 140
Bayer, Otto 86
Beethoven, Ludwid van 182–84
Beets 29
Belladonna 180
Bends 15
Benign prostate hypertrophy 73
Benzoquinones 33
Beta-carotene 114–15, 198
Betacyanin 30
Black cohosh 41, 135–36
Black, Joseph 230
Black tea 109–10, 177
Blood 48
Bombardier beetle 33
Bordeaux mixture 69
Boyer, Robert 16, 91
Brain sludge 79
Brandy 199
Breast cancer 120
Bromelain 163–64
Bromide 113
Bromine 112–13
Brown eggs 110–11
Brown sugar 204
Bubonic plague 123–24
Butyric acid 193

Calabar bean 143
Calcium carbonate 67, 101, 108,
    213–14
Calcium hydrogen carbonate
    213–14
Calcium oxalate 125–26
Calcium oxide 137
Calcium sulfate 108
Calories 219–20
Calcium sulfide 88
Camellia sinensis 109
Camphor 221–23

Candles 190–91
Cap pistols 39
Carbamide peroxide 11
Carbon dioxide 24, 68, 101, 185,
    206–7
Carbon monoxide 208
Carmine 9
Carotenoids 58–59
Carrothers, Wallace 85
Carrots 84, 114–15
Carson, Rachel 118
Cassava 174–75
Catechins 177
Caves 101–2
Cetirizine 80–81
Chaconine 194–95
Chamber pots 55
Chasteberry 46
Chemurgy 91
Cherries 171
Chloramine 158–59
Chlorine 69, 113, 158–59
Chlorofluorocarbons 12
Chlorogenic acid 217
Chlorophyll 197, 198
Chloroquine 144
Chocolate 26, 117–18
Cholesterol 78–79
Christmas pudding 216–17
Civet cat 111
CLA 168–70
Cleft palate 48
Clostridium difficile 201
Cochineal 9
Cockleburs 178
Cockroaches 35
Coffee 111–12, 213–14
Coins 229
Collagen 163
Collins, Francis 167
Collodion 226
Color-blind 20

Coniine 207
Conjugated linoleic acid 168
Copper 28, 104
Copper sulfate 68–69
Corfam 86
Corn 228–29, 231
Corn syrup 197
Cortés, Hernán 9
Cortisol 72
Cottage cheese 80
Courtois, Bernard 154–55
Crookes, William 42–43
Cryolite 141
Curie, Pierre and Marie 148
Cyanide 77, 175, 208
Cyclopentane 13
Cytosine 121

Dactylopius coccus 9
Dalton, John 19–21
Daltons 231–32
Davy, Sir Humphrey 156
DDT 118–21
De Chardonnet, Count Hilaire
    225–26
De Mestral, George 178
De Osorio, Lady Ana 75
Delessert, Jules Paul Benjamin 30
Detergents 206
Diabetes 157
Diamond Kit 60
Dibutyl phthalate 32
Dichloroethane 153
Dinosaurs 101
Dioxins 208
DNA 93, 167, 121, 231
Downy mildew fungus 69
Drew, Richard G. 44
Dripless candles 190
Durians 53–54
Dynamite 186

Earwigs 37
Ecstasy 222
Egg(s) 149, 220
Emollients 78
Emu 82
Emu eggs 84
Emu oil 81–83
Enzymes 206
Epidermal growth factors 14
Eraser 56–58
Ergot fungus 71
Erythromycin 228
Estragole 84
Estrogen 135–36
Ethanol 17
Ethyl–2-cyanoacrylate 202–3
Ethylene 181
Evolution 34

Fahlberg, Constantin 150–51
False fingernails 153–54
Fay, Annie Eva 43
Fermentation 64
Ferric tannate 176
Fish 165
Fizz Keeper 24
Flea 123
Floating soap 187–88
Flu 141
Ford, Henry 91
Formic acid 66–67
Fox sisters 43
Frankenstein 73–75
Freezing 146–147, 192
Fugu 173

Galvani, Luigi 74
Galvanism 74–75
Gasoline 129
Gay-Lussac, Joseph 156
Gel candles 18–19
Gelatin 124, 197

Genetic modification 232
Genetically modified foods 231
Gingivitis 80, 81
Global-warming 12
Glow-in-the-dark 88–89
Glucose 209, 228
Glycyrrhizic acid 72
Goiter 156, 209–10
Gold 140, 230
Grecian bend 15
Greek fire 18
Green tea 109
Guanine 121

Hall, Charles Martin 141
Hal the Healer 59
Hancock, Thomas 61
Hanukkah 115–16
Helicobacter pylori 201
Hell 67
Hemocyanin 104
Hemoglobin 48, 103–4
Henry's Law 15, 24
Hercules 182
Héroult, Paul Louis Toussaint 141
Hippocrates 14, 40, 144–45
Hirsch, Dr. Alan 26
Hofmann, Albert 71
Honey 201
Hydrazine(s) 52, 84
Hydrochloric acid 189
Hydrocortisone 81
Hydrofluorocarbons 13
Hydrogen cyanide 77
Hydrogen peroxide 11–12, 48–49,
    127, 176
Hydrogen sulfide 220–21
Hydrogenation 87
Hydrotherapy 51

Ice cream 11
Ink eradicators 176

Instant coffee 149
Insulin 158
International units (IU) 103
Iodine 155–56, 209–10
Iron 176–77, 217
Iron oxide 94
Isobutene 13
Isopropanol 191
Ivory 188

Jelly beans 124–25
Jesuits 76
Jewelry 132

Kidney stones 106, 125–26
Kissing 162–63
Konzo 175
Kopi Luwak coffee 111
Kraft, J. L. 215
Krazy glue 202–3

Lake Nyos 68
Lapis solaris 88
Lead 182–83
Lead acetate 126–27
Leblanc, Nicolas 187
Lemon juice 176–77
Licorice 73
Licorice root 72
Lidocaine 35–36
Lightbulb 168
Limestone 102
Linamarin 175
Lind, James 74
Lobster 58
LSD 71–72
"Lucy in the Sky with Diamonds"
    72
Luminous paints 88
Lycopene 70
Lydia Pinkham's Vegetable Com-
    pound 40–41
Lysergic acid diethylamide 71

Maccabees 115–16
Maggots 128, 172
Magnesium 94
Maillard reaction 117
Malaria 75, 143
Manganese dioxide 39
Marble 108
Margarine 54, 87–88
Marseilles vinegar 134
Marshmallow 196–97
McConnell, James 62–63
Mège-Mouriès, Hippolyte 87–88
Menopause 135
Mercury 165–66
Methane 161
Methanol 151–52
Methyl mercaptan 220–21
Methylcyclopentadienyl manganese
    tricarbonyl (MMT) 129
Methylmercury 166
Microwave 201–2
Mildews 193
Milk 197–99
Miss Piggy 85–86
Moir, E. W. 16
Molecule 25
Monomethylhydrazine 130
Montezuma 9
Montgolfier, Jacques-Étienne and
    Joseph-Michel 98–99
Morning glory 71
Morphine 131–32, 180
Moses 184, 223–24
Mother of vinegar 17
Müller, Paul Hermann 119
Myristicin 84

Natural vitamins 102
Nelson, Admiral 199–200
New York Stress Tabs 41–42
Nexia Biotechnologies 155
Nickel 132

Nitric acid 109
Nitrogen tetroxide 130
Nitroglycerin 186–87
Nobel, Alfred 186–87
Nobel Prize 186–87
Nooth, Dr. John 65
Nylon 85

Octopus 103
Olive oil 115
Ololiuqui 71
Onions 218
Oolong tea 110
Opium 131, 160
Ouzo 99–100
Oxygen 52, 185

Padimate O 93
Pamukkale 66–67
Pandemic 141–42
Paracelsus 84, 134
Paregoric 41
Pariza, Dr. Michael 169
Pasteur, Louis 63–64, 114, 224–25
Patent medicine 40
PBSA 93
Perkin, William Henry 160–61
Perón, Eva 122
pH 105–6
Phenol 41
Phenolphthalein 48–50
Philosopher's stone 133–34
Physostigmine 143
Phytoestrogens 135
Pig carcasses 172
Pinkham, Lydia 40
Plexiglas 153
Poison 207
Poly(acrylonitrile-butadiene-
    styrene) 153
Polyacrylate 153
Polyacrylonitrile 77

Polyethylene 212–13
Polyphenoloxidase 216–17
Polyphenols 110
Polyurethane 12, 85–87
Polyvinyl chloride 32, 57
Poppy 131
Porphyrins 111
Potassium 72–73
Potassium chlorate 39
Potassium cyanide 77
Potassium nitrate 155
Potato clock 27
Potatoes 194–95, 216
Priestley, Joseph 56, 65, 98
Processed cheese 215
Procter and Gamble 187
Progesterone 46
Propanethial-s-oxide 218–19
Propolis 100–101
Prostate 72
Prostate cancer 70
Protein 80, 154–55, 231
Pseudomonas 193
Puffer fish 173

Quinacrine 144
Quinine 41, 75, 160, 143–44

Radiometer 42–44
Radium 148
Rayon 226
Red ant 147
Red clover 135
Red phosphorus 39
Red tide 184–85
Respiration 77
Resveratrol 164
Retina 21
Revere, Paul 108
Riboflavin 198
Roebling, Washington A. 16
Romeo 42, 159

Roosters 107–8
Rubber 56–57, 61–62
Rubber cement 44

Saccharin 150–51
Safrole 85
Sal ammoniac 29
Salicylic acid 105
Salt 122–23, 209–10
Sassafras 85
Saturated fats 54
Scotch tape 44–45
Serotonin 72, 217
Serturner, Friedrich 131
Shakespeare 159–60
Shelley, Mary 73–75
Shrimp 58
Silicone dioxide 136
Silkworms 225
Simpson, O. J. 48–49
Skywriting 188
Smelling salts 28
Smoke detectors 211
Snails 103
Snake 14
Soap 138
Soda water 65
Sodium 50
Sodium bicarbonate 193, 210
Sodium erythorbate 52–53
Sodium hydroxide 50, 138
Sodium metabisulfite 90
Sodium monohydrogen
    phosphate 215
Sodium oxide 137
Sodium stearate 190
Sodium sulfite 51–52
Sodium thiosulfate 210
Solanine 194–95
SOS pads 204
Soy 135
Soybeans 91

Space shuttle 94
SPF 92
Spider silk 155
Spiders 103
Spiritualism 43
St. Pierre, Dr. Jean 60
Starch 228
StarLink corn 232
Steakhouses 195
Stormy petrel 191
Streptomycin 227
Styrene-ethylene/styrene-
    butadiene 19
Sugar 30, 64, 196, 201, 205
Sulfur 109
Sulfur dioxide 89–91, 108–9
Sulfuric acid 77, 108–9
Surface tension 191
Surfactants 203
Synthetic gas 152
Synthetic vitamins 102

Taco sauce 48
Tannic acid 176
Tea 176–77
Temple of Apollo 67
Tetrodotoxin 173
Thermite 93
Thymine 121
Thyroid disease 156
Titanium dioxide 93, 188
Titanium tetrachloride 188–89
Tooth-whitening 11
Transmutation 96
Trihalomethane 159
Tristearin 138
Tungsten 168

Tyre 182
Tyrian purple 182

Unsaturated fats 54
Uranium 96
Urea 117
Urine 30, 105–6

Vampire bat 216
Vampires 215
Van Houten, Conrad 117
Velcro 178–79
Venter, J. Craig 167
Verdigris 174
Vicks VapoRub 142
Vinegar 17, 79, 158
Vitamin B12 139–40
Vitamin C 102
Vitamin D 93
Vitamin E 102–3
Vitex agnus-castus 46
Volcanoes 89

Waksman, Dr. Selman 227
Water intoxication 50
Watermelon 70
White eggs 110
Windshield washer fluid 151–52
Wine 17, 164–65
Wohler, Friedrich 117
Wooden plates 195
World Cup 85–87
Worms 62

Zeolite 32
Zinc 28
Zinc oxide 93
Zyklon B 77

ABOUT THE AUTHOR

Dr. Joe Schwarcz is the director of McGill University's Office for Science and Society. He is the author of *That's the Way the Cookie Crumbles*, *The Genie in the Bottle*, and *Radar, Hula Hoops and Playful Pigs*. In addition to his weekly radio program and newspaper column, "Dr. Joe" makes regular television appearances and is a frequent public lecturer. He has received a number of honors in recognition of his work, including a 2003 Independent Publishers Book Award for best science book of the year and the American Chemical Society's Stack-Grady Award for interpreting science to the public.